Imagine a World without Monarch Butterflies

Awakening to the Hazards of Genetically Altered Foods

Alex Jack

Foreword by Congressman Dennis J. Kucinich

One Peaceful World Press
Becket, Massachusetts

Special thanks to Michio and Aveline Kushi, Congressman Kucinich, my mother, my wife, my daughter, my associates at One Peaceful World, my colleagues and students at the Kushi Institute, friends on the EarthSave—Boston Internet discussion group, and the Monarch butterflies who are showing us the way.

Imagine a World without Monarch Butterflies:
Awakening to the Hazards of Genetically Altered Foods

One Peaceful World Press
P.O. Box 10
Leland Road
Becket, MA 01223
U.S.A.

Toll-Free in the U.S. 1-888-322-4095
Telephone (413) 623-2322
Fax (413) 623-6042

First Edition: 2000

10 9 8 7 6 5 4 3 2 1

ISBN 1–882984–39–0
Printed in U.S.A.

Contents

"And the process was an all-or-nothing one; either you failed to modify at all, or else you modified the whole way."
—Aldous Huxley, BRAVE NEW WORLD

Ode to the Monarch Butterfly

"But most of all I shall remember the Monarchs, that unhurried westward drift of one small winged form after another, each drawn by some invisible force. . . . Did they return?"
—Rachel Carson, LETTER, 1963

"The eggs—fluted golden ovals—darken in a few days, then give forth comma-sized caterpillars. These consume milkweed flesh until they burst their skins. . . . In less than a month the caterpillars have grown as large as a child's little finger. After the fifth molt comes the jade jewelbox, the gold-nippled smooth green chrysalis. From this inch-long pellet will come a butterfly whose wings span three or four inches. Once they break free of their confining chrysalides, then dry in full turgor, they are capable of rising, soaring, and gliding better than any kite. On the first morning breeze they take wing, rise on the thermals, level out, float and pump higher, higher until, hundreds of perhaps thousands of feet in the air, they begin the masterful glide that can carry them many miles."
—Robert Michael Pyle, CHASING MONARCHS

"It's easy to get mesmerized watching the Monarchs glide overhead, with the sun shining through their wings. . . . They are silent, beautiful, fragile; they are harmless and clean; they are determined; they are graceful; they stalk nothing; they are ingenious chemists; they are a symbol of innocence; they are the first butterfly we learn to call by name. Like the imagination, they dart from one sunlit spot to another. To the Mexicans, who call them las palomas, *they are the souls of children who died during the past year, fluttering on their way to heaven."*—Diane Ackerman, THE RAREST OF THE RARE

"All around the country, farmers are about to finish sowing millions of acres of a genetically altered form of corn that protects itself from pests by producing a toxin in its tissues. But researchers report today that this increasingly popular transgenic plant, thought to be harmless to nonpest insects, products a wind-borne pollen that can kill Monarch butterflies—a species that claims the Corn Belt as the heart of its breeding range."—NEW YORK TIMES, *May 21, 1999*

"Most corn pollen remains within the cornfield and Monarch larvae can choose to avoid feeding on Bt pollen by feeding on the underside of leaves or on other milkweed leaves with little or no Bt pollen."
—Monsanto's official response to evidence that gene-altered corn was killing Monarch larvae

Foreword

If we are what we eat, then we must know what is in our food, so we may know what we are to become. *Imagine a World without Monarch Butterflies* is a thought-provoking journey into the omnipresence of genetically engineered products in our society. Alex Jack has put together a powerful compendium of well-researched information which is essential reading for anyone who wants to understand the awesome challenge which genetically altered foods presents to all of humanity.

Throughout Alex Jack's explication of the broad effects of genetic engineering on our society, he establishes *the public interest* as being a transcendent, irreducible imperative in matters of peace, life, health, safety, and the environment. As the public becomes better informed of the effects and risks of genetic alteration of food, an enlightened public can pursue choices which will truly be in their best interest.

Alex Jack demonstrates that in a democracy, the inherent rights of the people must be continually reclaimed and articulated in order to be assured. Government has a moral responsibility to ensure the purity and safety of the food supply. We cannot abdicate this responsibility to global corporations whose goals may be limited to profit-orientation.

Trade laws, tariffs, and sanctions are being used as economic weapons to advance the interests of genetically altered products. Alex Jack's work proves why we must protest provisions of the General Agreement on Tariffs and Trade (GATT) which are conducive to the mindless spread of this technology which is used to undermine labor, health, and safety standards, as well as to attack the sacred principle of national sovereignty. The World Trade Organization (WTO) would assume hegemony over our rights to self-determination, to our choice of food, and to our right to know what is in the food we eat. Alex Jack's book provides us with practical, holistic, and spiritual means to respond.

We all must eat so that we may live. But food is more than fuel. Food choice is a very personal act. Partaking of food expresses cultural and ethnic affirmation, religious affiliation, ethical choices, political, economic, and social orientation. Genetically engineered food represents a challenge to freedom of choice and freedom of expression. It touches something very deep in all of us—that in our striving to consume that we not become less than that which we consume. Alex Jack understands that it is a high expression of humanity to question, and to resist, the homogenizing power of global economic structures.

Thanks to the work of individuals like Alex Jack, I am preparing to introduce a bill in the House of Representatives which will require the labelling of all genetically engineered foods. It is imperative that consumers be able to know what foods are genetically engineered.

Can we change this system which has forcefully been put upon us, without our knowledge, without our permission? The human heart is very powerful and the human soul is unconquerable. One person who holds fast to his or her ideals and strives for justice in the face of seemingly impossible odds can stir the transformation of social, political, and economic structures. This is why markets will change. This is why genetically engineered products will be labelled. We must always believe in our ability to change things. One person can truly make a difference. Alex Jack has made a big difference with *Imagine a World without Monarch Butterflies*.

Dennis J. Kucinich
U.S. House of Representatives
Washington, D.C.

Dennis J. Kucinich is the U.S. Representative from the 10th District in Ohio. He served as Mayor of Cleveland and in Congress has been active on issues related to food, health, and the environment.

1
Introduction
Ending the War on Nature

"If you can look into the seeds of time
And say which grain will grow and which will not."
—Shakespeare, MACBETH

As the 21st century begins, the world food supply is undergoing rapid transformation. For the first time, human beings are eating foods that have not developed naturally—foods whose genetic structures have been changed in ways that millions of years of natural evolution could never achieve. Genetically modified foods (also known as genetically altered and genetically engineered foods) have moved invisibly into the marketplace. Today a majority of items in American supermarkets and restaurants—and possibly even natural foods stores—include GM ingredients. No long-term studies have been done on the impact of these new foods on health and the environment. No labeling is required by the U.S. government, and the ordinary individual or family has no way of knowing what they are buying in the store, eating at the restaurant, or even growing in their garden. At the present time:

- An estimated 90 million acres in the U.S. are planted with GM crops, constituting about one fourth of the total farmland
- This includes about 55% of soybeans, 35% of corn, 40% of cotton, and 5% of potatoes
- 30% of American dairy cows are in herds injected with GM bovine growth hormone (BGH) (which is banned in Europe and Canada)
- 50-60% of processed foods in the U.S. contain GM foods or ingredients, especially soy and corn derivatives. This includes margarine, mayonnaise, salad dressing, shortening, bread, and baked goods
- Most meat, chicken, eggs, and other animal products are produced from livestock fed genetically altered corn, soybeans, and cotton

Modern science has contributed many benefits to society. New technologies such as genetic engineering have several positive dimensions. For example, DNA screening is now widely used to analyze blood and bodily fluids and has resulted in the release of scores of individuals convicted of crimes they did not commit. Similarly, paternity suits and ancestral bloodlines are now being convincingly established on the basis of genetic testing. Anthropology, archeology, and several other fields may benefit.

Applied to food production and agriculture, however, genetic engineering is fraught with risks. Because no long-term studies have been done, we have no way of gauging their impact on personal health, social health, and planetary health. Preliminary short-term studies suggest potentially serious consequences to humans, plants, and animals. Unlike many consumer products, such as automobiles, toasters, or even drugs and medications, genetically altered crops cannot be recalled if they are found to be unsafe. They are out there forever, multiplying, mutating, and spreading novel genes, viruses, and toxins, and overturning 4 billion years of natural evolution. Their effects are irreversible.

Most of the rest of the world, especially Europe, has begun to control and limit this new technology. Sensitized by the epidemic of mad cow disease, the British Medical Association called for a moratorium on the introduction of GM crops and foods, pending comprehensive studies of their impact on health and the environment. The Sierra Club, National Wildlife Federation, Consumer's Union, and other groups have also called for a halt. Scientists are warning about an increase in allergies, immune-deficiency diseases, cancer, and other disorders, as well as peril to insects, birds, and mammals.

Virtually alone among nations, America has ignored the warnings and forged recklessly ahead to redesign the world's flora and fauna. But that changed almost overnight when Cornell University researchers reported that the pollen from GM corn could migrate to adjacent milkweed plants and kill the larvae of Monarch butterflies. The peril to the majestic orange and black creatures—a symbol of beauty, perseverance, and hope and widely regarded as America's national insect—served as a wake up call to the nation. The national press began more serious, in-depth coverage of the subject, especially the looming trade war between America and the European Union over this issue.

Who would have thought—ten years ago, five years ago, even a year ago—that an international food fight would become the central millennial drama of our time, eclipsing the contest between communism and capitalism, East vs. West, or the protracted racial, religious, and ethnic conflicts that have characterized most of the 20th century? Yet many of the leading issues of the last hundred years are converging with the introduction of genetically altered crops, foods, and products. Decentralization vs. globalization. Human rights vs. property rights. Public disclosure vs. corporate secrecy. An educated citizenry vs. a scientific elite. A dynamic, energetic view of the world vs. a fixed, mechanistic view.

Since 1992, when the FDA declared that engineered foods would be regulated no differently from regular foods, the biofoods industry has transformed American agriculture with virtually no regulation, oversight, or public awareness of the hazards involved. The "don't know, don't test" mindset that prevailed through most of the 1990s has now peaked. As public health, scientific, environmental, and religious organizations in the U.S. weigh in on this issue, fueled by grassroots activists, organic farmers, and natural foods manufacturers and consumers, the biotech industry is in full retreat. In late 1999, Monsanto (the leading biotech company) abandoned its plan to introduce "Terminator" seeds in the Third World—GM seeds that go sterile and cannot be replanted, thereby overturning thousands of years of natural agriculture and the saving of seeds by farmers. Robert Shapiro, Monsanto's CEO, admitted, "We forgot to listen. We have irritated and antagonized more people than we have persuaded. Our confidence in biotechnology has been widely seen as arrogance and condescension." In this rapidly changing social, economic, and political climate, some form of consumer labeling in the U.S. is inevitable. As with Big Tobacco, product liability and billion-dollar lawsuits against genetically engineered foods and products will follow. Ultimately, GM food makes for bad science, bad business, and bad eating. Even the animals won't eat it (see "Animal Intuition," p. 35)!

Winning a victory on labeling is not enough, however, as GM crops pose a serious and present danger to organic crops through "genetic drift." Biotechnology is just one manifestation of a trend toward increasing artificialization of the modern world. Even before genetic engineering was developed, an estimated 97% of all native species of grains, beans, vegetables, and fruits in America disappeared in the 20th century, driven to extinction by monoculture, hybrid seeds, jet transportation, and modern economies of scale. In the last 25 years, according to new USDA Food Composition Tables posted on the Internet, the nutrient content of most garden vegetables has declined between 25% and 50%. Beyond the use of biotechnology for food, there are medical and pharmaceutical applications, including xenotransplants (implanting an organ from one species into another), cloning, new reproductive technologies, and other artificial methods and procedures that also threaten the natural biological and spiritual evolution of our species.

Imagine a World Without Monarch Butterflies is part of the impassioned grassroots movement to alert the public to the magnitude and scope of the danger, as well as to present a straightforward, dispassionate summary of the evidence currently available on the health and environmental hazards of biotechnology. Although, by definition, no long-term studies have been done, there is a surprising amount of short-term research that raises serious questions about the viability of new genetic technologies. As much as possible, original scientific and medical journals and journalistic sources are cited using the format in *Let Food Be Thy Medicine: 750 Scientific Studies on the Personal and Planetary Benefits of Whole Foods* (One Peaceful World Press, Third Edition, 1999). Statements by organizations and individuals are also presented, including Michio Kushi, Jeremy Rifkin, John Fagan, Mae-Wan Ho, Marc Lappé, Arpad Pusztai, Vandana Shiva, Ronnie Cummins, Nina Moliver, and other leading authors, scientists, and community activists who are on the forefront of this issue. The resource section at the end lists books, organizations, and web sites devoted to GM food issues.

Over the last generation, our country has undergone a tremendous health revolution. The importance of a balanced diet based on whole grains, vegetables, and other fresh foods has been widely recognized. It will take dramatic, concerted action to protect freedom of choice, end the war on nature, and ensure the health of America and the planet as a whole.

Alex Jack
Becket, Massachusetts

2
Consumer's Guide to Genetically Modified Foods

"The concern is that a label would be seen as a stigma, like a skull and crossbones."
—Carl Feldbaum, president of the Biotechnology Industry Organization, *Washington Post*, August 15, 1999

Scores of genetically modified foods and products have been introduced around the world. Those currently available include:

- United States (50+ foods in total)
- Canada (30 foods)
- Japan (22 varieties of 6 crops)
- European Union(9 foods)
- Argentina (3 foods)
- Mexico (3 foods)
- Australia (2 crops—cotton, carnations)
- Brazil (1 food)
- South Africa (1 food)
- China (1 crop—cotton)

Selected U.S. Companies Using GM Foods

- Crisco (shortening)
- Frito, Dorito, Tostito (corn chips)
- Green Giant (harvest burger)
- Isomil and ProSobee (soy formula)
- Kellogg's (corn flakes)
- Kraft (salad dressings)
- McDonald's (french fries)
- Nabicso (sundry)
- Nestle (chocolate, dairy)
- Old El Paso (taco shells)
- Ovaltine (malt beverage)
- Parkey (margarine)
- Pillsbury (sundry)
- Procter & Gamble (sundry)
- Quaker Mills (sundry)
- Wesson (vegetable oils)

Sources: *Consumer Reports*, September 1999, pp. 41-46; *New York Times*, September 8, 1999; and other published reports.

Corn and Other Grain Products

Corn is the only genetically modified grain currently on the American market. An estimated 35% of the U.S. corn crop, including corn for both animal feed and human consumption, is now engineered. GM corn contains Bt, a bacteria that releases a toxic protein that is designed to kill the corn borer and other organisms that can damage the crop, but the plant's pollen can migrate to adjacent milkweed plants and kill the larvae of Monarch butterflies. Many processed food products in the supermarket and natural foods store contain corn syrup, cornstarch, corn dextrose, corn oil, corn flour, or other corn product that may be genetically altered.

In Japan, scientists announced that they have produced an altered form of rice that contains three times more dietary iron than conventional rice. The high-iron rice is made using by inserting a soybean gene that produces a protein called ferritin into the rice plant DNA. Meanwhile, Swiss and German researchers are developing a rice engineered to have a vitamin A derivative with genes spliced from a daffodil and from a bacterium.

Genetically Modified Corn Products

- Corn on the Cob
- Popcorn
- Corn Tortillas
- Grits
- Polenta
- Corn Syrup
- Corn Fructose
- Corn Starch
- Corn Dextrose
- Corn Oil
- Corn Flour
- Other Corn Products

Genetically Modified Grain Products in Development

- Rice
- Wheat
- Barley

Genetically Modified Processed Foods and Products with Corn Ingredients

- Corn Chips
- Cookies
- Candies and Gum
- Bread
- Cereals
- Pickles
- Margarine
- Alcohol
- Enriched Flours and Pastas
- Salad Dressings
- Vanilla

Soybeans and Soy Products

Soybeans are the only altered beans currently available in the U.S. and commonly are spliced with genes that help make them resistant to herbicides or alter their oil content. The *Journal of Medicinal Foods* reported the results of an independent study showing that GM soybeans have from 12 to 14% less phytoestrogens than normal. These are naturally occurring substances that help protect against cancer and heart disease.

About 50% of the American soybean crop is genetically engineered. GM soybeans, like most GM foods in the U.S., are produced by Monsanto, a large biotech company headquartered in St. Louis. They are sold commercially as Roundup Ready Soybeans because they are designed to withstand the application of Roundup, the herbicide which Monsanto also manufactures and sells to farmers. Since Japan imports 86% of its soybeans from America, many processed soy foods sent back to the U.S. may contain GM soy. In one independent spot test conducted by the *New York Times*, a majority of soy products tested positive for GM ingredients. Soybean oil constitutes 80% of the vegetable oil consumed in America and is used in margarine, salad dressings, mayonnaise, shortening, and other common foods. Many other foods and products contain soy products or derivatives such as lecithin, soy protein, and soy flour.

Genetically Modified Soy Products

- Soybeans
- Tofu
- Tempeh
- Soymilk
- Miso
- Shoyu
- Tamari
- Lecithin
- Soybean Oil
- Soy Flour
- Soy Protein
- Soy Isolates
- Genistein
- Other Soy Products and Derivatives

Processed Foods and Products with Genetically Modified Soy Ingredients

- Soy hotdogs
- Soy burgers
- Soy cheese
- Soy yogurt
- Dairy Ice Cream
- Frozen Yogurt
- Sauces and Dressings
- Candies, Cookies, Chocolate
- Bread and Baked Goods
- Breakfast Cereals
- Peanut Butter
- Protein Powder
- Infant Formula
- Shampoo
- Cosmetics
- Other Foods and Products

Potatoes, Tomatoes, and Other Vegetables

GM potatoes (such as the Burbank Russet) commonly have built in pesticides and are used to make french fries, mashed potatoes, baked potatoes, potato chips, and other products. They are also used to make potato starch and potato flour which is found in many processed foods. In laboratory studies in Scotland, GM potatoes fed to rats resulted in stunted growth and damage to major organs, including kidney, spleen, thymus, and stomach. To its credit, McCain Foods USA, the world's largest potato company, has not embraced the new technology and requires farmers to declare if they are using GM potatoes. Several varieties of tomatoes on the market are altered and include spliced organisms that may withstand herbicide applications. One variety of engineered squash is also now available, with many others vegetables expected to be introduced in the next few years.

Genetically Modified Vegetables
- Potatoes
- Tomatoes (regular and cherry)
- Yellow Squash
- Red-Headed Chicory (Radicchio)

Genetically Modified Vegetables in Development
- Peppers
- Cucumber
- Peas
- Broccoli
- Carrots
- Cauliflower
- Lettuce
- Sweet Potatoes
- Beets
- Other Vegetables

Genetically Modified Processed Foods Containing Potato or Tomato Ingredients
- French Fries
- Mashed Potatoes
- Baked Potatoes
- Potato Chips
- Potato Soup
- Tomato Sauce
- Tomato Soup
- Tomato Purée
- Lasagna
- Pizza
- Italian Foods
- Mexican Foods
- Other Foods and Products

Milk and Dairy Products

An estimated 30% of the cows in the U.S. are in herds given Bovine Growth Hormone (BGH), a genetically engineered growth hormone that increases yields. In medical studies BGH has been linked with cancer and it increases mastitis and other diseases in dairy cows. The European Union and most recently Canada have banned the use of BGH. Genetically engineered enzymes are also used in cheese production, and Consumers Union reported that 60% of all hard cheese products on American shelves are made with an engineered form of rennet. Animal feed (including corn, soybeans, and cotton) commonly includes GM ingredients, so that almost all non-organic dairy products includes engineered components.

Genetically Modified Dairy

- Milk
- Butter
- Cream
- Sour Cream
- Whey
- Buttermilk
- Ice Cream
- Yogurt
- Dairy or Soy Cheese made with Chymosis or Chymax, (GA Rennet)

Processed Foods Containing Genetically Modified Dairy Ingredients

- Whipped Cream
- Cottage Cheese
- Milk Shakes
- Cocoa
- Candies
- Cookies
- Bread
- Cake Mixes
- Sauces and Dressings
- Soups
- Other Products with Dairy

Meat, Poultry, Fish, and Other Animal Foods

Genetically engineered cattle, sheep, chickens, and fish are in development but their meat, milk, eggs, or other products have not yet been approved for human consumption. (Cloned beef is merketed in small quantities in Japan and the United States.) However, with the exception of animal food produced from organically grown grains or other natural foods, almost all meat, diary, poultry, and factory-bred fish in the United States are raised on feed that is genetically altered or contains GM ingredients. Up to 90% of America's total corn and soybean production goes to feed livestock, and from one third to one half of these crops are genetically altered. Moreover, 90 to 95% of soy meal, including the outer hulls of the beans, used in human foods are recycled in animal feed. Cottonseed oil and cotton byproducts are also added to silage, up to 50% in some cases, and fed to livestock. About 40% of the cotton grown in the U.S. is genetically engineered.

Animal Foods Commonly Made with Genetically Modified Feed (Corn, Soy, and Cotton)
- Beef, including Hamburger, Steak, etc.
- Pork, Ham, Hot Dogs
- Lamb
- Chicken, Eggs, Turkey, and Other Poultry
- Factory-fed Trout, Salmon, and Other Fish

Genetically Modified Fish and Seafood in Development
- Abalone
- Atlantic Salmon
- Catfish
- Prawns
- Trout

Fruits and Juices

Papayas are the only fruit currently engineered, but bananas, grapes, strawberries, and many others are expected to appear in the next few years. Fruit drinks at the present time may contain GM corn syrup and corn dextrose. Dried fruit is commonly sprayed with an oil derived from soybeans that may be GM. This includes raisins, sultanas, currants, dates, and dried fruit in breakfast cereal.

Gentically Modified Fruits
- Papaya

Genetically Modified Fruits in Development
- Apples
- Grapes
- Strawberries
- Pineapples
- Bananas
- Melons
- Other Fruits

Canola, Cotton, Peanut, and Other Oils, Seeds, and Nuts

About 60% of the canola oil produced and sold in North America is GM. Canola is an increasingly popular oil in restaurants and institutional cooking because of its polyunsaturated quality, light texture, and mild taste. It is widely used in processed foods and products. Along with soybeans, corn, and canola oil, cotton is one of the four major engineered crops in America. An estimated 50% of all cotton grown in the US is GM. In addition to clothing, linens, and other fabric, its derivatives, especially cottonseed oil, are used in manufacturing chips, peanut butter, cookies, crackers, and other processed foods. GM peanuts just entered the market and are used in peanut oil for cooking and peanut butter.

Genetically Modified Oils, Seeds, and Nuts
- Canola Oil
- Cottonseed Oil
- Peanuts and Peanut Oil

Genetically Modified Oils, Seeds, and Nuts in Development
- Chestnuts
- Sunflower Seeds
- Walnuts
- Other Oils, Seeds, and Nuts

Processed Foods and Products Containing Genetically Modified Oils or Cotton
- Chips
- Cookies
- Crackers
- Margarine
- Fried Foods
- Sauces and Dressings
- Soups
- Baked Goods
- Peanut Butter
- Soaps
- Detergents
- Cottons (Jeans, T-Shirts, etc.)
- Linens
- Other Fabrics
- Other Products Containing Canola, Cotton, or Peanuts

Vitamins and Supplements

Several vitamin supplements especially vitamin C, is made with corn fructose which may be genetically altered.

Genetically Modified Supplements
- Vitamin C

Enzymes

Enzymes are proteins that accelerate biological processes. They are used widely in the food industry to make beer, bread and baked goods, sugar, dairy foods, and other products. Because they are not considered foods, enzymes are not required to be labeled on products. Now new genetically engineered enzymes has been introduced. They also are unlabeled, and the government does not require that manufacturers notify the FDA. The following GM enzymes are known to have been introduced:

Genetically Modified Enzymes

- Ampha Amylase (White Sugar, Corn Syrup, Honey)
- Aspartic (Cheese)
- Chymosis (Cheese)
- Novamyl (Bread and Baked Goods)
- Pullulanase (High Fructose Corn Syrup)

Foods Commonly Made with Enzymes

- Bread and Baked Goods
- Beer and Wine
- Dairy Products
- Fruit Juices
- Oils
- Sugar

How Can You Tell What Foods Are Safe?

In the absence of mandatory labeling, there is no way to know whether a food contains GM ingredients without testing it in a genetic laboratory. GM foods tend to be more uniform, bigger, and more blemish free than usual foods, and in some cases GM perishable foods are reported to last for months without spoiling. As a general rule, smaller, irregular, less shiny, and faster ripening and faster spoiling foods contain a better balance of nutrients and energy and are safer to eat, even though they may not win a beauty contest.

Safety Issues Related to Organic Foods

Whole unprocessed foods that are organically certified are generally grown from natural seed and do not contain genetically altered ingredients. Processed or packaged organic foods can in some states contain up to 5% non-organic ingredients, which could be GM. The biotech industry lobbied extensively for its GM seeds to be considered organic, but after opposition by the organic farming community and the natural foods movement USDA Secretary Dan Glickman pledged that new national organic standards soon scheduled to be released will, by definition, not include GM components.

Unfortunately, this does not mean that organic food is necessarily GM free. Some organic fields, for example, those growing organic corn, have become contaminated by "genetic drift," pollen blown by the wind from nearby genetically altered corn fields. A major organic farm in Britain was reportedly decertified as organic because of such contamination.

Another major concern is that GM corn, cotton, and other crops will produce new strains of Bt-resistant organisms that will spread to organic farms. Naturally occurring Bt—a much less toxic variety than engineered Bt—is the most widely used pesticide on organic farms. The organic foods community is worried that its industry could be destroyed if natural Bt is rendered ineffective as a result of GM Bt.

Commenting on the drift of GM modified corn and contamination of organic fertilizer, Gary Anson, an organic farmer in Calhoun, Missouri, commented, "It's coming at me from every direction. I've got nowhere to hide."

Geneticist Bill Beavis, a researcher at the National Center for Genome Resources and a supporter of genetic engineering, conceded that he is worried about the health effects of GM crops and the risks of introducing them into the environment. He said that the challenge of keeping them separate from organic crops is baffling because organic crops could still be contaminated in a cooperative's grain elevator or mill. "It's just virtually impossible to segregate. We'd have to change our whole agricultural system to do that [and] it would interfere with the freedom of farmers to do as they please."

Margit Kaltenekker, a certifier for the Organic Crop Improvement Association, reported that many organic farmers are spending hundreds of dollars per field for genetic tests to prove their crops are not contaminated. "It's tough for them," she said. "They can do everything right and still be ruined by the guy a mile away."

Chicken feed, commonly used as fertilizer on organic food, may also contain remnants of feed made from gene-spliced corn.

Meanwhile, in Europe regulators for the European Union are discussing whether to set permissible limits for genetic contamination on non-GA foods, including organic food, because they question whether genetic pollution can be controlled. An allowable threshold of up to 1/2 to 1 percent GM material has been proposed to protect organic and sustainable farmers whose crops may unintentionally contain minute amounts of altered ingredients.

Sources: Mothers for Natural Law; Scot Canon, "Missouri Organic Farmers Struggle to Keep Crops Chemical-Free," *Kansas City Star*, August 24, 1999; and Ronnie Cummins, "Hazards of Genetically Engineered Foods and Crops," Campaign for Food Safety, August 1999.

• **Consumer Group Calls for Organic Standards and Crop Liability:** In autumn 1999, Consumers Union called for national standards that prohibited GM foods or ingredients from being labeled organic. It also called on the U.S. government to require a comprehensive review of the safety of altered foods before they are marketed, a policy to hold the biotech industry liable for economic or health damages resulting from GM crops, and mandatory labeling. "Consumers have a fundamental right to know what they eat," the nation's largest consumer organization stated.

Source: *Consumer Reports*, September 1999, pp. 41-46.

3
The Health Hazards of Genetically Modified Foods

"The agency is not aware of any information showing that foods derived by these new methods differ from other foods in any meaningful or uniform way."
—U.S. Food and Drug Administration (FDA)

"The Europeans have an absolute fear, unfounded by any scientific basis, of accepting these products."
—Stuart Eizenstat, nominee for undersecretary of the U.S. Treasury Department, testifying before the U.S. Senate, June 29, 1999

1. Increased Allergic Reactions and Lung Problems

• **Altered Plants Potentially Lethal:** In 1996, researchers at the University of Nebraska reported that rapeseed and soybeans, modified with genes from Brazil nuts to boost their low cysteine and methionine content, produced proteins that could result in extreme, potentially deadly allergic reactions in people sensitive to the nuts. "Since genetic engineers mix genes from a wide variety of species," noted Dr. Rebecca J. Goldburg of the Environmental Defense Fund, "other genetically engineered foods may cause similar health problems. People who are allergic to one type of food may suddenly find they are allergic to many more." Studies indicate that 2 percent of adults and 8 percent of children have food allergies.

Source: J. Nordlee et al., "Identification of a Brazil-Nut Allergen in Transgenic Soybeans," *New England Journal of Medicine* 334:688-92, 1996.

• **Altered Pollen May Affect the Lungs:** Blown by the wind, pollen from GM crops can spread across a large area of the U.S., affecting people and animals as well as the food on which the pollen settles. Risk assessment of GM crops "has ignored the impact of such GM pollen or has falsely claimed that the impact is negligible without even looking at data or background studies," according to Joe Cummins, professor emeritus of genetics at the University of Western Ontario. "GM pollen, particularly the pollen taken into the airway creates a unique threat that should be evaluated now and should have been evaluated prior to the release of the GM crops." Cummins said that modified Bt, for example, may prove to produce "allergens or to show toxicity different in the behavior of the protein in the gut." In addition to the lungs, he said "GM pollen is likely to open an extended range of allergy and may cause autoimmune disease or other kinds of toxicity."

Source: Joe Cummins, "Pollen Bystander Effects," June 13, 1999.

• **70% of Farmworkers Develop Allergies to Bt:** One month after harvesting vegetables sprayed with ordinary non-engineered Bt, half the crop pickers and handlers tested positive for allergies and within 3 months that share climbed to 70%, according to a report in *Environmental Health Perspectives*. In a letter to *Science News*, Mary-Howell Martens expressed concern that "many, many more people will be exposed to the Bt toxin and will likely be sensitized by eating crops engineered with the gene for the toxin."

Source: "Bt Is as Bt Does," *Science News*, September 4, 1999.

• **Inadequate Testing:** Commenting on allergies, Dr. Arpad Pusztai, a geneticist and former senior researcher of the Rowett Research Institute in Aberdeen, Scotland, explained in an interview that lack of predictability concerning GM foods is worrisome for people with food allergies. "These people can only live their lives on the basis that they know which foods to avoid. Biotech companies claim they test for "known allergens" like peanuts. But there are thousands of other foods that can cause serious allergies but which are not classed as known allergens. On top of this, there may be new toxins or allergens in GM foods that are not spotted because they are not looked for."

Source:: "Why I Cannot Remain Silent," *GM-Free* 1:2, August/September 1999.

2. Increased Toxicity and Stress on the Liver

• **Roundup May Affect the Liver:** The main ingredient in Monsanto's herbicide Roundup is glyphosate, a pesticide that has produced increased bile acids, alkaline phosphatase and alanine aminotransferase, and other changes in laboratory animals. Glyphosate can also inhibit the monooxygenases required by mammals to detoxify other chemicals. Genetically altered Roundup Ready soybeans are commonly used with Roundup to protect crops from the harmful effects of glyphosate. However, scientists speculate that glyphosate enters the human food chain through contaminated soybeans hulls mixed into animal feed. "Heavy meat eaters might have exposure from meat and animal byproducts that become contaminated in the course of feedlot programs using Roundup Ready soybeans," explains crop researcher Marc Lappé. "This is because livestock can be fed soybean hulls, and the bacteria that thrive in the intestinal tract metabolize glyphosate into a more fat soluble and toxic amine that could, in theory, accumulate in their body tissues."

Source: Marc Lappé, Ph.D. and Britt Bailey, *Against the Grain* (Common Courage Press, 1998), pp. 62-65.

3. Higher Risk of Herpes

• **Surgeon's Daughter Stricken:** GM soymilk triggered a herpes-related virus in the daughter of a senior surgeon, according to testimony presented to the British government. The surgeon, a women who wished to remain anonymous to protect her daughter, stated, "I want the Government to look into this because I saw the change in my daughter. As soon as she was taken off the GM milk, her health dramatically improved. I, and my GP, have not found any other reasons why she became ill."

Tests showed that the child is not allergic to soymilk, which her mother began feeding her in February, 1998, when she was just a year old because she had developed allergies to dairy products. The girl immediately began to develop large cold sores that did not heal. She drank about 4 pints of the GM milk every day. Three large, weeping sores developed on her face. When the soymilk was cut down to half a pint a day, the sores cleared up overnight.

Source: "GM Soya Milk Gives Children Herpes, Senior Surgeon Tells the Government," *Sunday Telegraph*, London, U.K., August 1, 1999.

4. Introduction of Foreign Proteins in the Body

• **EPA Official Speaks Out:** Observing that foreign proteins that have never before been part of the human food chain will soon be consumed in large amounts, Suzanne Wuerthele, a risk assessor at the U.S. Environmental Protection Agency (EPA) stated, "It took us 60 years to realize that DDT might have oestrogenic activities and affect humans, but we are now being asked to believe that everything is OK with GM foods because we haven't seen any dead bodies yet."

Source: Declan Butler and Tony Reichhardt, "Long-Term Effect of GM Crops Serves Up Food for Thought," *Nature* 398:651, 1999.

5. Altered or Reduced Nutrition

• **20% Less Protein in GM Potatoes:** In laboratory studies in Scotland, a researcher reported that GM potatoes had 20 percent less protein than normal.

Source: Brian Tokar, "Butterfly Experiment Highlights Biotech Hazards," *Food & Water Journal*, Summer 1999.

• **Novel Nutritional Effects:** "Certainly, the new protein product of the introduced gene, usually an enzyme, that detoxifies the herbicide or its metabolites, presents an additional component of the plant which may alter the bioavailability of essential nutrients or become a toxic or quasi-toxic component of the plant itself," asserts Marc Lappé, an expert in experimental pathology and director of the Center for Ethics and Toxics. "A further concern is that the gene product survives in the kernel or bean of the plant, adding unspecified amino acids and, possibly, new toxic byproducts to the commodity." He called for thorough testing before food or animal feed containing new proteinacous material is released.

Source: Marc Lappé, Ph.D. and Britt Bailey, *Against the Grain* (Common Courage Press, 1998), pp. 142-144.

6. Damage to Stomach, Intestines, and Blood

• **Rats Suffer Damage to Major Organs:** Scottish researchers reported that GNA, an insect and worm resistance factor (or lectin) found in GM potatoes, could cause widespread changes in the stomach and intestines of rats fed a diet of raw or boiled GM potatoes for only 10 days. A related study found that GNA also binds strongly to human white blood cells.

Source: Dr. Arpad Pusztai and Stanley Ewen, *Lancet* 354:1353-1355, 1999; and Brian Fenton et al., *Lancet* 354:1314-1316, 1999.

7. Impairment of the Immune Function

• **Foreign DNA and RNA in Genetically Altered Food May Provoke Immunological Reactions:** According to researcher Sharyn Martin, Ph.D., the autoimmune disease SLE, rheumatoid arthritis, and immune complex glomerulonephritis may be enhanced from exposure to foreign DNA fragments and nucleoproteins introduced by GM foods. Type III complex mediated hypersensitivity reactions, she warned, can affect the skin, producing edema and erythema of the lungs, pigeon fanciers disease, and pulmonary aspergillosis.

Commenting on her study, the Physicians and Scientists for the Responsible Application of Science and Technology observed, "Studies have shown that DNA is not fully digested in the gut. Large fragments, even of the size of a gene have been found to survive digestion and can be taken up into the bloodstream. In genetic engineering, a number of genes are used that have not existed in human food. The consequences of eating such DNA is therefore unpredictable. This article warns for the possibility that fragments of transgenic DNA might provoke chronic allergic or autoimmune disorders including such that cause serious disabling or even lethal disease."

Source: Dr. S. Marin, "Immunological Reactions to DNA and RNA," PSRAST, July 2, 1999.

8. Increased Risk of Cancer

• **Main Herbicide in GM Crops Linked to Cancer:** Food with residues of glyphosate, the herbicide widely used in genetically altered crops, may increase the incidence of certain cancers.

Source: L. Hardell and M. Erikkson, "A Case-Control Study of non-Hodgkin's Lymphoma and Exposure to Pesticides," *Cancer* 86(6), 1999.

• **New Cancers Possible:** Britain's Chief Medical Officer, Liam Donaldson, and Chief Scientific Adviser, Sir Robert May, told ministers of the government to set up a special panel to examine whether GM food could produce "fetal abnormalities, new cancers, and effects on the human immune system." The report recommended that the U.K. set up a health monitoring unit similar to the one that discovered a link between mad cow disease and human CJD.

Scientists including Dr. Michael Antoniou of Guy's Hospital in London have warned that GM food could lead to new allergies, cancers, and other human illnesses because of "the disruption of our natural genetic order." "The reasons why we can't be specific about the health consequences of GM food is that we don't know enough. Each genetic engineering event holds its own dangers. You could have acute toxicity or something that sneaks up over many years. Any of these things are possible."

Source: *The Independent*, London, U.K., May 2, 1999.

• **Threat of Malignancy:** Professor Mae-Wan Ho of the U.K. Open University Department of Biology warns, "Genetic engineering bypasses conventional breeding by using artificially constructed parasitic genetic elements, including viruses, as inside cells. These vectors slot themselves into the host genome. The insertion of foreign genes into the host genome has long been known to have many harmful and fatal effects, including cancer of the organism."

Source: Mothers for Natural Law.

• **BGH Increases Cancer Risk:** In a review of the evidence linking dairy and breast cancer, researchers with the Physicians Committee for Responsible Medicine contend that the FDA's approval of BHG, genetically engineered bovine growth hormone, was based on faulty assumptions and studies. While the FDA concluded that it did not differ chemically from natural BGH, studies indicate that BGH differs by 1 to 9 animo acids. The FDA also assumed that BGH is not orally active in humans and that its activity is destroyed during pasteurization, both of which have been contradicted by subsequent studies.

"A substantial body of medical evidence provides possible mechanisms by which milk may promote breast cancer," the physicians conclude. "(1) IGF-1 and estrogens are present in all milk in micromolar to nanomolar concentrations; (2) IGF-1 is not destroyed during milk pasteurization; (3) IGF-1 has been shown to stimulate or initiate growth of human breast cancer cells; (4) IGF-1 acts synergistically with estrogens, which increase its effects even at nonomolar concentrations; (5) BGH increases IGF-1 levels in milk; (6) IGF-1 and BGH can possibly be absorbed intact from the GI tract; (7) IGF-1 can exert local mitogenic tissue effects and be cleaved to exert local mitogenic tissue effects." While BGH has not been considered to be a danger because subsequent increase in bovine milk IGF-1 levels are within the "normal range" based on untreated cows and human breast milk, the physicians assert that the "normal range" could be carcinogenic when milk is ingested regularly over a lifetime. They conclude that milk produced with BGH may increase the risk of cancer.

Source: J. L. Outwater et al., "Dairy Products and Breast Cancer: The IGF-1, Estrogen, and BGH Hypothesis," *Medical Hypotheses* 48:453-61, 1997.

9. Increased Risk of Heart Disease

• **Reduced Phytoestrogens:** Scientists reported that GM soybeans yield 12-14% less phytoestrogens than usual. Phytoestrogens are beneficial components associated with reduced risk of heart disease and certain cancers.

Source: "Alterations in Clinically Important Phytoestrogens in Genetically Modified Herbicide-tolerant Soybeans," *Journal of Medicinal Foods* 1:4, 1999.

10. Increased Risk of Diabetes

• **BGH May Contribute to Diabetes:** Recent studies have linked diabetes with cows' milk formula for babies and infants. Tests are needed to determine whether the addition of BGH to the national milk supply could accelerate this process, according to a correspondent in a noted science journal.

Source: Warren Geisler, "Another Diabetes Connection?" *Science News*, August 21, 1999.

11. Death and Disability

• **Lethal Food Supplement:** In 1989, a genetically engineered form of L-tryptophan, a common food supplement used as a sedative by people with esosinophilia myalgia syndrome (EMS), produced toxic contaminants. Thirty-seven people in the United States died, 1500 others were permanently disabled, and 5000 became very ill, before the product was recalled by the FDA. Lawsuits against the manufacturer, Showa Denko, Japan's third largest chemical company, have resulted in payments of over $2 billion in damages.

Source: A. N., Mayeno and G. J. Gleich, "Eosinophilia-myalgia Syndrome and Tryptophan Production: A Cautionary Tale," *Tibtech* 12:346-52, 1994.

12. Delayed Sexuality and Reproductive Development

• **Retarded Sexual Development:** Laboratory-bred, Bt-resistant pink bollworms took longer to develop into mature adult moths than larvae feeding on non-Bt cotton, according to researchers at the University of Arizona. The implication of this study is that strategies of planting refuges of non-GA cotton close to a GM variety to slow the spread of insects that develop resistance to a pest-killing toxin in GM cotton may not be viable.

"If the insects develop to sexual maturity at different speeds, interbreeding is less likely or impossible and so may well speed the development of resistance and reduce the benefits of the GM crop," Michael Crawley of Imperial College in Ascot, Britain, commented.

Because both types of moth mate within 3 days of hatching and males live only for about a week, it could mean that Bt-resistant bollworms would mate with others, thus ensuring the spread of Bt-resistant bollworms and defeating the strategy of the biotechnologists.

Source: "U.S. Study Raises New Questions about GM Crops," Reuters, August 5, 1999.

13. Increased Risk of Infertility

• The Scottish Crop Research Institute found in 1998 that female ladybugs that ate aphids that had eaten GM potatoes laid fewer eggs and lived only half as long as ordinary ladybugs.

Source: Nick Nuttall, *The Times*, London, July 13, 1998.

14. Increased Risk of Birth Defects and Developmental Disorders

• **Transgenic Cotton Produces Birth Defects:** Bromoxynil, a toxic herbicide used in some varieties of transgenic cotton and sold under the brand name Buctril, produces teratogenic (fetal) effects and is classified as a "possible" human carcinogen by the EPA. It is associated with increased incidence of liver tumors in laboratory studies, as well as defects in spine and skull, reduced fetal weight, and developmental disorders in fetuses. While cotton is not a direct food crop, cottonseed oil is widely used in commercial foods and cotton leavings and waste makes up about 50 percent of the silage used in animal foodstuffs.

Source: Marc Lappé, Ph.D. and Britt Bailey, *Against the Grain* (Common Courage Press, 1998), p. 44.

• **Glufosinate Linked to Birth Defects:** Glufonsinate, the principal herbicide in genetically altered soybeans, corn, and canola oil, can lead to birth defects. Pregnant women who consume food containing glufonisate have a higher risk of giving birth to children with birth defects, learning disabilities, and other abnormal behavior. Researchers found that fathers exposed to the herbicide also gave birth to defective children even if the mother was not exposed.

Source: A. Garcia et al., "Paternal Exposure to Pesticides and Congenital Malformations," *Scandianivan Journal of Work-Related and Environmental Health* 24:473-80, 1998.

15. Emergence of Antibiotic-Resistant Bacteria

• **Mad Cow Expert Warns about Altered Food:** Deadly new bacteria, viruses, prions, and other substances have emerged in recent years as a result of the overuse of pesticides, antibiotics, and other chemicals, as well as contaminated and rendered food.

Dr. Harash Narang, the microbiologist who first alerted the British government to the dangers of mad cow disease to human health, warns that genetically modified foods are "a recipe for disaster." "If you look at the simple principle of genetic modification it spells ecological disaster," the senior research associate at the University of Leeds stated. "There are no ways of quantifying the risks." He said GM crops may lead to new drug-resistant strains of deadly bacteria that are resistant to ampicillin and other antibiotics. "The solution is simply to ban the use of genetic modification in food," he stated.

Source: *The Journal*, Newcastle, U.K., July 13, 1999.

• **EU Considering Ban on Antibiotic-Resistant Marker Genes:** Gene-altered foods commonly include an antibiotic resistance marker gene (ARM) to help determine whether the principal gene inserted into the plant or host organism was successful. Some scientists worry that ARM genes will recombine with bacteria or other microorganisms in the environment or in the digestive tract of humans or animals which ingest them and lead to new strains of pneumonia, tuberculosis, staph, e-coli, salmonella, campylobacter, enterococci, and others. The European Union is reportedly considering a ban on all engineered foods that contain ARM genes.

Source: Ronnie Cummins, "Hazards of Genetically Engineered Foods and Crops," Campaign for Food Safety, August 1999.

• **Horizontal Transfer of Transgenic DNA:** Bacteria in the mouth and pharynx can take up and express transgenic DNA, including antibiotic resistance marker genes, confirming the ability of modified genetic material to spread by horizontal gene transfer. Altered organisms survived for up to 60 minutes. Meanwhile, the UK Ministry of Agirculture Fisheries and Food reported that modified DNA in samples of oil seed, linseeds, soybeans, and wheat were not readily broken down by most commercial processing methods. As a result, animal feeds may contain gene-sized DNA fragments that can be passed along to humans.

Sources: D. K. Mercer et al., "Fate of Free DNA and Transformation of the Oral Bacterium Streptococcus Gordonii DL1 by Plasmid DNA in NHuman Saliva," *Applied and Environmental Microbiology* 65:6-10, 1999; J. M. Forbes et al., "Effect of Feed Processing Conditions on DNA Fragmentation Section 5 - Scientific Report, UK MAFF, London, 1998.

• **Antibiotic Resistance Genes May Remain in the Intestines:** Dutch researchers reported that in a computer-controlled model of the stomach and intestines, designed to simulate the human digestive process, antibiotic-resistant genes in food could jump to bacteria in the gut. "The results show that DNA lingers in the intestine, and confirms that genetically modified bacteria can transfer their antibiotic-resistance genes to bacteria in the gut," researchers reported in *New Scientist*. Prior to the experiment, biotech scientists claimed that such transference could not happen because the modified DNA breads down so quickly, but the Dutch experiment showed that DNA from the bacteria had a half-life of six minutes in the large intestine. "This makes it available to transform cells," said Robert Havenaar, the designer of the artificial gut.

Source: "Study Casts Doubts on Genetically Modified Food," Reuters, January 28, 1999.

16. Emergence of New Viral Diseases

• **New Viral Epidemics Possible:** The Cauliflower Mosaic Virus (CaMV), the most common virus gene used in genetic engineering, may recombine with infecting genes yielding new viruses that may be more infectious than the natural viruses, cause serious diseases, and cross species borders. CaMV is used in Roundup Ready soy of Monsanto, the Bt-Maize of Novartis, GM cotton and various varieties of GM canola. Describing experiments on GM crops, Dr. Angela Ryan, molecular biologist at Open University, U.K., observed, "The results show that the CaMV promoter is very likely to recombine with other DNA in the host genome, including dormant viral DNA, as well as with other viruses in the host cell. Transgenic lines containing CaMV promoters, which includes practically all that have been released, are therefore prone to instability due to rearrangements, and also have the potential to create new viruses or other invasive genetic elements. Such elements cannot be contained or controlled once their have entered the wider environment."

Source: Jaan Suurkula, M.D., "The Virus Hazard," Physicians and Scientists for a Responsible Application of Science and Technology, July 31, 1999.

• **New Viruses Akin to Hepatitis B and HIV:** Dr. Joe Cummins, professor emeritus of genetics at the University of Western Ontario, has warned about the potential for creating new viruses from GM crops. "It has been shown in the laboratory that genetic recombination will create virulent new viruses from such constructions. Certainly the widely used cauliflower mosaic virus is a potentially dangerous gene. It is a pararetrovirus meaning that it multiplies by making DNA from RNA messages. It is very similar to the Hepatitis B virus and related to HIV. Modified viruses could cause famine by destroying crops or cause human and animal diseases of tremendous power."

Source: Mothers for Natural Law.

• **Doctors and Lawyers Call for Ban on GM Animal Organs:** Doctors and Lawyers for Responsible Medicine called for a ban on transplants using genetically engineered animal organs. DLRM president Dr. Andre Menache, an official with the Israeli Ministry of Health, warned that pigs, a common source for human organ transplants, carry many viruses that are potentially lethal. At a London press conference, the alliance warned that previously unknown viruses, as deadly as AIDS, could pass from modified animal organs into their human hosts.

Source: BBC News, June 10, 1999.

• **GM Foods May Carry Vaccines:** Around the world, scientists are seeking to incorporate vaccines and drugs into genetically engineered foods. Malaria vaccine, for example, is reportedly being inserted in potatoes in South America. In the Netherlands, researchers are producing genetically altered plants whose nectar could be used to make honey containing vaccines or drugs. Scientists at the Center for Plant Breeding and Reproduction Research in Wageningen added genes for several drugs to the plants to produce a "healing nectar." According to *New Scientist*, the researchers found a way to activate a genetic switch, or promoter, in the plant so that the drugs are made only in the nectar. "Once the plants are fully grown and begin producing nectar, bees will be unleashed on them to produce honey that the researchers hope will contain the vaccine," the magazine noted. The Dutch researchers are also growing altered petunias to produce a vaccine against parvovirus, a viral disease that kills dogs.

Source: "GM Honey Could Help the Medicine Go Down," Reuters, June 23, 1999.

4
The Environmental Hazards of Genetically Modified Foods

"The problems that people feel about . . . GM organisms are not based on scientific results. Rather, it has to do with an irrational and collective fear."—William Daley, U.S. Commerce Secretary, Le Monde, September 15, 1999

"I should have known what fruit would spring from such a seed."—Bryon, Childe Harold's Pilgrimage

1. Decline of Butterflies and Other Insects

• **GM Corn Kills Monarch Larvae:** Scientists at Cornell University reported that genetically modified corn could kill Monarch butterflies. They found that 44% of Monarch caterpillars that ate milkweed leaves dusted with pollen from GM corn laced with GM Bt (*Bacillus thuringiensis*) died in four days, while none of the caterpillars that dined on real corn died. The surviving larvae experienced a weight loss of 60 percent.

Normally, Bt bacteria produce an inactive toxin that becomes activated only when absorbed into the highly alkaline digestive systems of the corn borer and other pest. The GM Bt crops produce up to 20 times as much toxin in its activated form in every plant tissue throughout the life cycle of the plant.

The researchers concluded that their results "have potentially profound implications for the conservation of monarch butterflies" and more research on the environmental risks of GM crops is essential.

In a follow-up study in an actual cornfield, entomologists at Iowa State University report that a year of mortality data confirm that the threat to the Monarch is real.

Source: John E. Losey, "Transgenic Pollen Harms Monarch Larvae," *Nature* 399:214, 1999 and "Seeds of Change," *Consumer Reports*, September 1999, p. 41.

• **Coalition Formed to Protect Butterflies:** In a letter to the EPA, a coalition of environmental organizations called on the government to restrict the plantings of Bt corn and strengthen programs for identifying and assessing the environmental risks of GM crops which it regulates. The letter, sent by the Sierra Club, National Wildlife Federation, Union of Concerned Scientists, Natural Resources Defense Council, Arizona-Sonora Desert Museum, and Defenders of Wildlife, singled out the threat to Monarch butterflies:

"EPA failed to assess risks of toxic pollen to nontarget Lepidoptera. In documents supporting the approval of Bt corn, the agency neither mentioned nor assessed the likelihood that pollen from transgenic corn would have an impact on monarchs or any other nontarget Lepidopterian species. Of special concern is the failure to evaluate impacts on the 18 moths and butterflies listed as threatened or endangered under the federal Endangered Species Act." The letter went on to ask the EPA not to renew the registration of Bt corn varieties that expire in either 2000 or 2001 and to require 40 to 80 foot borders of non-Bt corn around engineered varieties, thereby reducing the amount of toxic pollen carried beyond the field to plants that moths and butterflies eat.

Source: "Scientists, Environmentalists Urge EPA to Protect Butterflies: First Joint Petition by Major Green Groups on Gene-Altered Crops," Union of Concern Scientists, August 11, 1999

• **Lacewings at Peril:** The Swiss Federal Research Station for Agroecology and Agriculture reported that the death rate of green lacewings rose by one and a half times after they ingested corn borers eating engineered corn laced with GM Bt.

Source: "Angelica Hilbeck, *Environmental Entomology*, April 1998.

• **Bees Lose Sense of Smell:** European researchers reported that honeybees fed proteins from altered rapeseed "had trouble learning to distinguish between the smells of flowers" and died earlier than bees not exposed.

Source: "Seeds of Change," *Consumer Reports*, September 1999, p. 41.

• **Horizontal Gene Transfer among Species:** Genetically altered organisms could transfer novel genes and viral vectors to other species. Horizontal gene transfer may even take place in the digestive systems of protozoa, nematodes, the larvae of insects, and other organisms that live in the soil.

Source: K. Harding, "The Potential for Horizontal Gene Transfer within the Environment," *Agro Food Ind. Hi-Tech* 7:31-35, 1996.

2. Decline of Birds and Other Wildlife

• **Bird Decline:** Responding to public concern and the Royal Society for the Protection of Birds which fears biotechnology will accelerate the already sharp decline in bird populations, the British Government announced a four-year trial in which GM crops will be planted alongside ordinary crops in Britain and biologists will monitor weeds, wildlife, and insects in and around the fields for four years. Earlier, English Nature, the Government's wildlife adviser, urged a halt on the planting of gene-altered plants.

Source: Nick Nuttall, *The Times*, London, July 13, 1998.

3. Decline of Fish and Rise of Superfish

• **Transgenic Salmon Threaten Natural Species:** Genetically engineered fish under development for commercial production in America and China could displace natural strains within several generations if they are released or escape into the wild. English Nature, the conservation advisor to the British Government, warned that it would oppose the release of such fish in British waters unless they were made infertile. The group cited research studies showing that a growth gene injected into salmon eggs can result in an increased growth rate of up to 50 times in the wild.

A Purdue University study was also cited that found transgenic male salmon enjoyed a 400% breeding advantage because of their larger size over native Atlantic Salmon. The researchers concluded that in a population of 100,000 natural fish, the addition of 1 transgenic male would result in 50% altered fish by the sixteenth generation. Another study found that Channel catfish given salmon growth promoter genes by scientists from Auburn University and Stanford University avoided predators better than regular fish and could result in a strain of "superfish" that could upset the balance of nature. Carp, tilapia, the Japanese medaka, and several other varieties are currently being tested in biotech experiments.

Source: Charles Clover, "GM 'Superfish" Face Ban in British Waters, BBC, August 9, 1999.

4. Increased Disease and Suffering among Animals

• **Canada Bans BGH:** The Canadian government rejected the injection of BGH, a genetically engineered growth hormone, into dairy cows in Canada in 1999. After nine years of research, Health Canada ruled that BGH "presents an unacceptable threat to the safety of dairy cows." The government ministry said that BGH increased mastitis, an udder infection, by 25%, infertility by 18%, and lameness of 50%.

Source: "Monsanto Picks Up Its BGH and Goes Home," *Good Medicine*, Spring 1999, p. 22.

5. Contamination of Organic Crops

• **Cross-pollination of Organic Crops:** In the Netherlands, a natural foods company randomly tested organic corn chips it had imported from the U.S and discovered that they contained GM ingredients. Prima Terra, the manufacturer in the U.S., had made the chips from corn obtained from a certified organic farm in Texas. As it turned out, the Texas farm was situated in a region in which GM corn was planted. The pollen from the GM corn evidently migrated 6 miles to the organic fields, cross-pollinating with the organic corn seed. In the Netherlands, the importer was forced to destroy 87,000 batches of the corn chips, valued at $147,000, because they could no longer be sold as organic. The chips destroyed by the European distributor were found to have less than 0.1 percent GM corn.

Source: "GM Crops May Contaminate Nearby Organic Crops," Reuters, May 27, 1999.

• **Pollution of Organic Crops Inevitable:** Genetically altered crops will "inevitably" contaminate organic crops, according to a new research study funded by the British Government. Pollen and seed pollution cannot be entirely avoided, the report concluded, and advised that "acceptable levels" of contamination be set. The recommendation, made by the John Innes Center, one of Europe's chief GM research institutes, was strongly opposed by organic farmers. Currently, the U.K. requires that "buffer zones" of from 200 to 600 meters be set up around GM crops, but studies by the National Pollen Research Unit and the Soil Association found that GM pollen could be carried by bees up to three miles and even further by the wind. GM crops are now grown on four farm-size locations in England and on 140 small experimental sites in the countryside.

Source: "Debate Refused on GM Crops 'Risk,'" BBC, June 17, 1999.

• **Organic Community Unifies Against GM:** At the annual Scientific Conference of the International Federation of Organic Agriculture Movements (IFOAM), six hundred delegates from 60 countries unanimously supported a declaration against GM foods and agriculture. The declaration called on governments and regulatory agencies to immediately ban the use of genetic engineering since it involved negative and irreversible impacts on the environment, release of organisms into nature that cannot be recalled, elimination of the right of choice for farmers and consumers, violation of the fundamental property rights of farmers and endangering their economic independence, practices contrary to the principles of sustainable agriculture, and unacceptable threats to human health.

Source: Dr. Michael W. Fox, "Will Genetically Engineered Crops Mean Contaminated and Toxic Food, Bodies, and Ecosystems?" Humane Society of the United States, Washington, D.C. July 27, 1999.

• **Bt-Resistance May Spread as a Dominant Trait:** Environmentalists fear that genetically altered Bt may produce a toxin that will spread to the wild and breed a race of superpests. *Science* compounded this concern by reporting that European corn borers resistant to the toxin released by Bt may pass along this gene as a dominant, rather than a recessive trait. This would result in a quicker multiplication and spread of resistant species than previously expected. It could also pose a more serious threat to organic and low-chemical use farmers who use ordinary Bt as a natural pesticide.

Source: *Science* 284:965-67, 1999.

• **Environmental Organizations File Suite to Protect Organic Farming:** A coalition of environmental groups and organic farmers brought a suit in 1999 against the EPA to end its approval of genetically engineered crops grown with Bt. Greenpeace International, the Center for Food Safety, and the International Federation of Organic Agricultural Movements charged the EPA with "wanton destruction" of natural Bt, which it calls the "world's most important biological pesticide."

6. Contamination of Conventional Crops

• **Contamination of Ordinary Crops:** Swiss and German officials reported in 1999 that non-GA corn contained novel genes from a variety of corn genetically modified to be resistant to the corn borer. Because the plants concerned are not commercially available in GM form, it appears that they were contaminated by cross-pollination with GM varieties imported from the U.S. The Swiss immediately banned the import and trade of the contaminated corn and ordered the destruction of any seeds already sown.

Source: "Swiss Farmers Seek GM Compensation. Normal Maize Has Unapproved GM Genes," *Nature Biotechnology* 17: 629, 1999.

7. Contamination of Wild Plants

• **Altered Rapeseed Penetrates Wild Species:** Scientists reported that herbicide-resistant genes from an altered rapeseed plant (used to make canola oil) spread into a nearby stand of wild field mustard.

Source: James Kling, "Could Transgenic Supercrops One Day Breed Superweeds?" *Science* 274: 180-181, 1994.

• **Genetic Drift:** Research by the Soil Association in the U.K. found that more than 80 percent of rapeseed pollen, used in making canola oil, is carried by bees and bees can travel more than three miles. The wind can transport it much further. Scottish researchers found that even at sites 400 meters away from genetically modified plots, as many as 7% of seeds from non-GA rapeseed plants were herbicide resistant.

Sources: Friends of the Earth Press, April 14, 1999; *The Independent*, London, June 16, 1999.

• **Genetic Drift:** Danish scientists reported that a gene inserted in a domesticated crop plant could spread to the wild. In the test, oilseed rape, a common commercial source of canola oil, was genetically altered to make the plant more tolerant of the herbicide glufosinate. When planted with closely related weeds, hybrids were produced that carried the new characteristic. Altogether, 42 percent of the second generation plants carried the new traits. Dr. Gus A. de Zoeten, an expert on genetic crops, commented, "I don't think this means creating an uncontrollable, 'Frankenstein plant' is more likely, but it shows that genes released into the environment eventually will escape, in essence creating a form of contamination."

Source: T. Mikkkelsen et al., "The Risk of Crop Transgene Spread," *Nature* 380:31, 1996.

8. Deformed and Stunted Crops

• **Deformed Cotton:** In 1997, Monsanto paid $5 million in damages to farmers in the Mississippi Delta when Roundup Ready cotton failed. The GM cotton was overrun with bollworms after application of Roundup. Farmers in Mississippi, Arkansas, Tennessee, and Louisiana lost much of their crop when deformed bolls developed and fell off the plants.

Source: Stan Grossfeld, "Genetic Engineering Debate Shifting to America," *Boston Globe*, September 23, 1998.

9. Decline of Soil Fertility

• **Land Infertility:** In an Oregon State University study, a GM bacterium developed to aid in the production of ethanol unexpectedly produced residues which rendered the land in which it was planted infertile. New corn crops planted on this soil grew three inches tall and fell over and died.

Source: H. R. Hill, "OSU Study Finds Genetic Altering of Bacterium Upsets Natural Order," *The Oregonian*, August 8, 1994.

• **Bt May Remain Active for Thousands of Years:** Genetically altered Bt toxin does not break down as rapidly in some soils as natural Bt and may remain after harvest or in compost. The cumulative effect could be to poison many beneficial insects and organisms necessary for the fertility and health of the soil. Further, the "naked" DNA remaining in decaying cells in the soil may remain biologically active for thousands of years. Such naked DNA may be eaten by mice and be inherited by offspring or be eliminated in feces and eaten by other animals or wildlife.

Sources: C. Crecchio and G. Stotzky, "Insecticidal Activity and Biodegradation of the Toxin from Bacillus Thuringiensis Subsp. Kurstaki Bound to Humic Acids from Soil," *Soil Biology and Biochemistry* 30:463-70, 1998 and T. Traavik, "Too Early May Be Too Late: Ecological Risks Associated with the Use of Naked DNA as a Biological Tool for Research, Production, and Therapy," *Research Report for DNA*, 1999. Trondheim, Norway, Directorate for Nature Management.

10. Spread of New Viruses

• **Pathogenic New Viruses:** A study reviewing 125 biogenetically engineered plant strains found that 3 percent of the altered plants produced pathogenic new viruses by exchanging genes with other plants. Corn, wheat, tomatoes, and other plants manipulated by scientists commonly carry up to five different viruses that can swap genetic material.

Source: A. E. Greene and R. F. Allison, "Recombination Between Viral RNA and Transgenic Plant Transcripts," *Science* 263:1423-25, 1994.

• **Viral Epidemic:** A genetically engineered virus escaped from quarantined research facilities in Australia and spread like wildfire through the rabbit population with disastrous results.

Source: I. Anderson, "Australia's Rabbits Face All-Out Viral Attack," *New Scientist*, September 7, 1996.

11. Blight, Epidemic, and Hunger

• **Lessons of the Irish Potato Famine:** Like potatoes that were initially welcomed as a blessing in Ireland in the 19th century, new genetically altered foods may fail and result in mass hunger and starvation, according to an article in the *Irish Times*. "The bulk of the Irish population became utterly dependent on that extraordinary wonder food, the lumper potato. For a long time, the humble spud supported a population that, though materially poor, was remarkably healthy by the standards of the time. And then a completely unforeseen natural occurrence, the arrival of potato blight, wiped out the crop and caused a traumatic transformation whose consequences are still with us," observed Fintan O'Toole. "The future of Irish agriculture would be much better served by taking a stance against GM crops. And, in the process, we would show some awareness of our own past."

Source: F. O'Toole, "Scientific Progress Is Blighted by GM Crops," *Irish Times*, Aug. 20, 1999.

• **GM Threatens the Gene Pool:** According to ecologists, biodiversity is essential to protecting crops from drought, blight, and other natural catastrophes. "Often the resulting broad-based gene pool proved essential to protect their fields from blights or other depredation," explains agricultural researcher Marc Lappé. "All of these organisms have coevolved by adapting to a changing environment. By losing plant varieties, plant growers are losing opportunities to breed for essential natural defenses."

Source: Marc Lappé and Britt Bailey, *Against the Grain* (Common Courage Press, 1998), p. 102.

12. Emergence of New Disease-Resistant Insects

• **Resistance 1000 Times Higher Than Expected:** The widespread use of pesticides, herbicides, antibiotics, and other chemicals in modern agriculture, food production, and medicine has generated fear of new drug-resistant microorganisms and other species. The National Academy of Sciences reported that 1.5 diamondback moth larvae out of every 1000 tested carry a gene resistant to genetically engineered Bt toxin, a level one thousand times higher than expected.

Source: B. E. Tabashnik et al., "One Gene in Diamondback Moth Confers Resistance to Four Bacillus thuringiensis Toxins," *Proceedings of the National Academy of Sciences* 94: 1640, 1997.

• **Beetle Predators Develop Resistance to Bt:** In Scotland, scientists report that aphids are capable of sequestering the toxin from Bt crops and transferring it to its lady beetle predators, in turn affecting reproduction and longevity of the beneficial beetles. "The potential of Bt toxins moving through food chains poses serious implications for natural biocontrol in agro-ecosystems."

Source: Miguel A. Altieri, "The Environmental Risks of Transgnic Crops: An Agroecological Assessment," *Pesticides and You*, Spring/Summer 1998.

13. Increased Pesticide Burden

• **Use of More Chemicals Forecast:** About two-thirds of all genetically altered crops are designed to be herbicide-resistant. The purpose of these herbicide-resistant crops is to allow farmers to spray as much pesticide and herbicides on their fields as they desire without damaging the crops. "Scientists estimate that herbicide-resistant crops planted around the globe will triple the amount of toxic broad-spectrum herbicides used in agriculture," reports environmental activist Ronnie Cummins. "These broad-spectrum herbicides are designed to literally kill everything green." As selective pressures create new, herbicide-resistant strains and Superweeds emerge, farmers will turn to ever increasing dosages and eventually strong, more toxic chemicals.

Source: Ronnie Cummins, "Hazards of Genetically Engineered Foods and Crops," Campaign for Food Safety, August 1999.

• **Special Tolerances for Roundup Approved:** The U.S. Environmental Protection Agency raised its level of permissible pesticide tolerances for Roundup residues in silage. "Use of herbicide tolerant crops virtually guarantees that beef, poultry, and pork will have higher contamination levels of selected pesticides than such livestock had previously," observed public health advocate Marc Lappé. He said that the special tolerances were authorized to increase use of Roundup Ready soybeans and that, as a result, consumers, especially children and infants, will be exposed to "a largely untested and different burden of pesticides than before" as the pesticide residues make their way up the human food chain.

Source: Marc Lappé and Britt Bailey, *Against the Grain* (Common Courage, 1998), p. 147.

14. Mutagenesis and Inheritance of Recessive Genes

• **'Silent Genes' May Lead to Mutagenesis:** Genetically engineered genes that code for enzymes may modify the metabolism of plants leading to an increase in toxins. "Mutagenesis by genes randomly inserted into the genome also raises the risk that 'silent' or lowly expressed genes might be switched on or upregulated with damaging results," according to a scientific report in *Nature*.

Source: Declan Butler and Tony Reichhardt, "Long-Term Effect of GM Crops Serves Up Food for Thought," *Nature* 398:651, 1999.

15. Bioinvasion and Irreversible Environmental Effects

• **Out of Control Spread:** A primary cause of species extinction is *bioinvasion*, where seeds, plants, insects, or microbes from a distant ecosystem are brought to new locale with devastating effects. The introduction of kudzu in the American South is a well known example. As Jeremy Rifkin observed, "Each new synthetic introduction is tantamount to playing ecological roulette. That is, while there is only a small chance of it triggering an environmental explosion, if it does, the consequences could be significant and irreversible." In 1997, 60,000 bags of rapeseed seeds, enough to plant 600,000 acres of land to produce canola oil, were recalled in western Canada after a gene unapproved for market unexpectedly turned up in the seeds.

Sources: Jeremy Rifkin, *The Biotech Century* (Tarcher/Putnam, 1998), p. 73; *Manitoba Cooperator*, April 4, 1997.

• **Animal Intuition:** On a four-month journey across the Corn Belt, a farm journalist found that cattle won't touch GM corn stubble, hogs won't eat their ration when GM crops are included, and raccoons romp by the dozen in the organic corn but leave the Bt fields untouched. One farmer described observing a herd of 40 deer "mowing down his tofu beans while across the road there isn't one doe eating on the Round-up Readies. . . Even the mice will move on down the line if given an alternative to these 'crops.' What is it that they know instinctively that most of us ignore? I have been traveling around with a bag of contaminated cob corn on the floor of my vehicle, and I have begun to think of it as if it was a bag of plutonium."

Source: Steven Sprinkel Yankton, "When the Corn Hits the Fan," *ACRES USA*, Sept. 18, 1999.

5
The Social and Cultural Hazards of Genetically Modified Foods

"We by art unteach what Nature taught."
—Dryden

1. Denial of Freedom of Choice

• **Consumers Cannot Make an Informed Selection:** Over the last generation, product safety and the "right to know" have been enshrined in American jurisprudence. Packaged foods carry nutritional labels, cigarettes bear warning labels, drugs and medications come with comprehensive descriptions of their contents, including benefits and risks, and potentially toxic or carcinogenic chemicals, building materials, and home furnishing must be clearly identified."Today, virtually all articles in commerce carry identification and labels of one kind or another," explains health researcher Marc Lappé. "For genetically engineered crops and their resultant byproducts to remain unlabeled flies in the face of both these precedents and common law practice." Without proper labeling, shoppers cannot make an informed choice about the foods they eat or feed their families.

Source: Marc Lappé, Ph.D. and Britt Bailey, *Against the Grain* (Common Courage Press, 1998), p. 129.

• **Majority of Natural Foods Test Positive for Altered Ingredients:** Most shoppers, and indeed most food merchandisers, including natural foods stores, are unaware that their food may contain genetically altered ingredients. In a test of several natural foods to see whether they contained altered ingredients, 10 of 11 products randomly selected by the *New York Times* tested positive in an independent genetic laboratory, including two soy-based infant formulas and three corn chips.

Source: "Revising a Bioengineering Test," *New York Times*, March 25, 1998.

• **Half Million Sign Petition Demanding Labeling:** A petition signed by nearly 500,000 Americans demanding the labeling of GM foods and products was submitted to the U.S. Congress, President Clinton, and several federal agencies, including the USDA, FDA, and EPA, on June 17, 1999. The petition was organized by Mothers for Natural Law in Fairfield, Iowa, an offshoot of the Transcendental Meditation movement.

Source: *Natural Law Party News* 3:2, 1999, p. 1.

• **Vast Majority of American Public Supports Labeling:** A *Time* survey in June, 1999 found that 81% of Americans want GM foods to be labeled. Other surveys found that about 60% of consumers would avoid buying GM foods if they were clearly labeled. The International Food Information Council reported that only 2 to 3% thought soybeans were altered.

Source: *Time*, June 00, 1999; *New York Times*, September 6, 1999.

2. Threat to Freedom of Speech and Other Civil Liberties

• **Organic Companies Sued for Labeling Products GM-Free:** The Food and Drug Administration (FDA) takes the position that GM foods are safe and comparable to ordinary foods. The agricultural biotech industry says that labeling a safe product would unnecessarily call attention to it, reduce consumer confidence, and adversely affect sales. Natural foods companies that have tried to implement "reverse labeling," letting the consumer know that GM ingredients are not included in their food, have been brought to court by the biotechnology companies for casting aspersions on their industry and sued for millions of dollars in damages.

Thirteen states where biotechnology companies are concentrated have passed "food disparagement" laws. These laws are widely considered unconstitutional, but the threat of lawsuits and the enormous legal expenses needed to challenge such suits have caused most natural foods companies to refrain from labeling their products "GA-free." In Vermont, for example, voters approved a referendum requiring the labeling of all dairy products containing BGH. In 1996, the state's labeling law was reversed by the U.S. Circuit Court of Appeals following an appeal by the Dairy Association with Monsanto as "a friend of the court." Ben & Jerry's, the ice cream manufacturer that has been leading a campaign to label their dairy products BGH-free, has been the subject of a lawsuit brought by the biotech industry. In 1998, Ben & Jerry's won the right in an Illinois lawsuit to keep a label on their products saying that it didn't include GM ingredients or BGH as long as it included a disclaimer saying the FDA considered BGH milk equivalent to ordinary milk.

Meanwhile, the biotech firms have been using their economic clout to stifle public debate on the issue. Monsanto has pressured several publishers to withdraw or tone down forthcoming books on the subject. J. Robert Hatherill, a research scientist at the University of California at Santa Barbara, had long passages cut out from his book, *Eat to Beat Cancer*, containing information on BGH. Vital Health Publishing of Bloomingdale, IL, canceled Marc Lappé's *Against the Grain* after receiving a letter from a Monsanto lawyer, who said he believed the manuscript, which he had not read, included false statements that would disparage Roundup, the herbicide made by Monsanto. In a front page article, David Richard, owner of Vital Health, told the *New York Times*, "I was scared. As soon as I told my insurance agent about the letter, he would not return any of my calls. I had no choice. I had to let go of the book."

In Ohio, where a food libel law as passed in 1996, the threat to free speech has also had a chilling effect. "When I give speeches, I look around and think, 'Does someone have a tape recorder?' said Laurel Hopwood, a nurse and Sierra Club volunteer who speaks to organizations about genetically altered food. "I'm even afraid to say 'This might be unsafe because I'm fearful I could get sued." She said that she and her friends were afraid to hand out brochures on GM foods on Earth Day because of the possibility of being sued.

Source: Melody Petersen, "Farmers' Right to Sue Grows, Raising Debate on Food Safety," *New York Times*, June 1, 1999.

• **British Scientist Fired:** Dr. Arpad Pusztai, a senior researcher at the Rowett Research Institute in Aberdeen, Scotland, was condemned by his superiors and forced to retire after his research showed that GM potatoes caused stunted growth, damage to the immune function, and major organ damage in laboratory experiments. Pusztai, a noted geneticist who published 270 scientific papers during his 35 years at Rowett, said he himself would not eat GM foods and said it was "very, very unfair to use our fellow citizens as guinea pigs." *Lancet*, Britain's leading medical journal, later published his study over the objection of the biotech industry.

Source: *Rachel's Environment & Health Weekly* #649, May 6, 1999; Dr. Arpad Pusztai and Stanley Ewen, *Lancet* 354:1353-1355, 1999.

3. Reduced Crop Yields and Economic Decline

• **Reduced Yields:** American farmers reported mixed success with GM crops. While the engineered seeds produce greater yields, farmers can lose money when commodity prices and insect infestations are low, according to a study by the National Center for Food and Agricultural Policy, a biotech industry research group based in Washington.

In 1997, corn growers made an extra $72 million by using GM seeds, but in 1998 after planting three times as much acreage lost $26 million when grain prices dropped and infestation levels plummeted. Cotton farmers using GM crops saved $72 million in 1998. Potato farmers have not planted much GM crops because they like to use an insecticide that kills more pests.

"An accurate assessment of the contribution of a new pest control technology would require a decade or more of actual field usage," the reported concluded.

Source: Philip Brasher, "Study Finds the Success of Bt-Crops Is Mixed," Associated Press, July 14, 1999; *Food Chemical News*, July 5, 1999, pp. 13-14.

• **Roundup Rounds Up Less Yields:** Testing conventional soybeans and GM Roundup Ready soybeans under similar growing conditions, the Arkansas Cooperative Extension Service reported that in 30 out of 38 comparisons in 1995 and 1996, the altered soybeans on average yielded 4.38 less bushels per acre than conventional soybeans.

Source: Marc Lappé, Ph.D. and Britt Bailey, *Against the Grain* (Common Courage Press, 1998), p. 84.

• **The Last Roundup?** In a review of 8,200 field trials, genetically altered Roundup Ready seeds produced fewer bushels of soybeans than conventional varieties. The study was performed by Dr. Charles Benbrook, former director of the Board on Agriculture at the National Academy of Sciences.

Source: Peter Rosset, "Why Genetically Altered Food Won't Conquer Hunger," *New York Times*, September 1, 1999.

4. Risky Investments and Reduced Earnings

• **World's Largest Bank Alerts Investors to Dump GMOs:** Deutsche Bank, the world's largest bank (based in Germany) posted an analysis on the Internet in the summer of 1999 signifying a major change in the market's view of genetically modified foods and products. Foreseeing a two-tier system taking hold, where farmers are paid a premium for conventional crops, the report predicted, "we see price premiums of high-value-added GMO seeds collapsing." "We predict that GMOs, once perceived as the driver of the bull case for this sector, will now be perceived as a pariah," the analysis continued. Characterizing the message as "a scary one" for farmers, manufacturers, and investors, Deutsche Bank asked: "Who has the liability if GMO pollen corrupts a non-GMO crop? Take that to the next logical step. If you plant GMO corn, and your neighbor plants non-GMO hybrids and pollen from your GMO crop drifts into his field, making his grain test positive for GMOs (even though he planted non-GMO hybrids), will your neighbor sue you? Will you be suing the seed company? Imagine the legal mess that could ensue if GMO = value destruction. Don't expect the food manufacturers and retailers to 'take a bullet for GMOs.'"

Forecasting "an earnings hit" and "an earnings nightmare," the bank reduced its rating on Pioneer Hi-Bred (makers of GM seed) to SELL from HOLD. It also forecast that Monsanto would be in financial trouble and wondered whether the acquisition of Delta & Pine Land, creators of Terminator Technology, would collapse. Overall, it likened GM food to the nuclear power industry.

Source: "Is Genetic Engineering in Agriculture Dead?," Deutsche Bank, July 12, 1999.

5. Increased Insurance Risks

• **Insurance Company Expresses Doubts:** A high insurance company official with Cigna International, an insurance giant, warned in an article in a trade publication that issuing insurance policies to genetic engineering companies was risky. "Our experience with asbestos, PCBs, and other 'miracle' products in the past should have warned us of the potential dangers of diving into issues before we have an adequate awareness of the exposures," Maunce Pullen stated.

Source: *Post*, May, 6, 1999.

6. Increased World Hunger and Poverty

• **Terminator Technology:** In 1998 the USDA and Delta and Pine Land Company, a Mississippi company and the largest cotton seed producer in the world, developed a new biotechnology called Technology Protection System that created sterile seed by programming a plant's DNA to kill its own embryos. The process, termed "Terminator Technology," can be applied to plants and seeds of all species. "The goal is to increase the value of proprietary seed owned by U.S. seed companies," explained Willard Phelps, USDA spokesman, "and to open up new markets in Second and Third World countries." Delta and Pine Land Company was subsequently purchased by Monsanto.

Source: *Mountain Farm Stewards*, May/June, 1998.

• **Bioserfdom:** Several billion people around the world live on subsistence farming and save seeds to plant the following year. The introduction of patented seeds, including Terminator Technology, could ultimately force many of these people to purchase increasingly expensive modified seeds and the pesticides and chemicals that accompany them. "If the trend is not stopped," warns environmental activist Ronnie Cummins, "the patenting of transgenic plants and food-producing animals will soon lead to universal 'bioserfdom' in which farmers will lease their plants and animals from biotech conglomerates such as Monsanto and pay royalties on seeds and offspring. Family and indigenous farmers will be driven off the land and consumers' food choices will be dictated by a cartel of transnational corporations. Rural communities will be devastated. Hundreds of millions of farmers and agricultural workers worldwide will lose their livelihoods."

Source: Ronnie Cummins, "Hazards of Genetically Engineered Foods and Crops," Campaign for Food Safety, August 1999.

• **Africans Oppose Terminator Technology:** In a conference on genetic engineering sponsored by the Food and Agricultural Organization of the United Nations (FAO) in 1998 in Rome, the African delegates issued a statement opposing Terminator Technology. The delegates strongly objected "that the image of the poor and hungry from our countries is being used by giant multinational corporations to push a technology that is neither safe, environmentally friendly, nor economically beneficial to us.

"Rather than developing technology that feeds the world, Monsanto uses genetic engineering to stop farmers from replanting seed and further develop their agricultural systems," the statement went on.

"In 'Let the Harvest Begin' the Europeans are asked to give an unconditional green light to gene technology so that chemical corporations such as Monsanto can start harvesting their profits from it. We do not believe that such companies or gene technologies will help our farmers to produce the food that is needed in the 21st century. On the contrary, we think it will destroy the diversity, the local knowledge and the sustainable agricultural systems that our farmers have developed for millennia and that it will thus undermine our capacity to feed ourselves.

"In particular, we will not accept the use of Terminator or other gene technologies that kill the capacity of our farmers to grow the food we need. We invite European citizens to stand in solidarity with Africa in resisting these gene technologies so that our diverse and natural harvests can continue and grow."

Source: *Mothers for Natural Law Newsletter*, August 1998.

• **Monsanto Terminates Terminator Technology:** In the face of worldwide opposition, Monsanto announced in late 1999 that it would abandon the commercial development of Terminator Technology. In a letter to the Rockefeller Foundation which opposed selling sterile seeds in the Third World, Robert Shapiro, Monsanto's CEO, said that his company had patents for "gene protection" and might develop other technologies in the future.

Source: "Monsanto Says It Won't Market Infertile Seeds," *New York Times*, October 5, 1999.

7. Patenting of Seeds and New Life Forms

• **Traditional African Plant Patented:** The University of California and a Korean pharmaceutical company received an American and international patents in 1993 for a genetically altered sweetener based on thaumatin, a traditional African plant. Native people have been using the exceptionally sweet plant as a traditional food. The goal of the scientists was to develop and obtain the proprietary rights for a low-calorie sweetener that could be genetically inserted in fruits and vegetables. "With the market for low-calorie sweeteners nearing a billion dollars in the U.S. market alone, thaumatin is likely to became a cash cow in the coming years," reported author and cultural critic Jeremy Rifkin. "Meanwhile, villagers in West Africa will not share in the good fortune, even though their ancestors are the real discoverers of thaumatin."

Source: Jeremy Rifkin, *The Biotech Century* (Tarcher/Putnam, 1998), p. 53.

• **Scientists Call for Ban on Patenting Genes:** Calling for a ban on patents on living organisms, cell lines, and genes, scientists associated with the Institute of Science in Society (ISIS) called for a comprehensive, independent public inquiry into the future of agriculture and food security, taking account of the full range of scientific findings as well as socioeconomic and ethical implications.

"The patenting of living organisms, cell lines, and genes under the Trade Related Intellectual Property Rights agreement are sanctioning acts of piracy of intellectual and genetic resources from Third World nations, and at the same time, increasing corporate monopoly on food production and distribution. Small farmers all over the world are being marginalized, threatening long term food security for all," the statement asserted.

"Governmental advisory committees lack sufficient representation from independent scientists not linked to the industry. The result is that an untried, inadequately researched technology has been rushed prematurely to the market, while existing scientific evidence of hazards are being downplayed, ignored, and even suppressed, and little independent research on risks are being carried out.

Source: World Scientists' Statement Calling for a Moratorium on GM Crops and Ban on Patents, Institute of Science in Society.

• **Biopiracy:** New biotech patents involve "acts of plagiarism of indigenous knowledge and biopiracy of plants (and animals) bred and used by local communities for millennia," according to an update of concerns prepared for the World Scientists' Statement on Biopatents.

All classes of new biotech patents should be rejected from inclusion in the World Trade Organization (WTO) and similar international agreements because "all involve biological processes not under the direct control of the scientist. They cannot be regarded as inventions, but expropriations from life." The "hit or miss technologies" associated with many of the new "inventions" are inherently hazardous to health and the environment. There is "no scientific basis" to support the patenting of genes and genomes "which are discoveries at best." They also create unnecessary suffering in animals and are "otherwise contrary to public order and morality."

Source: Mae-Wan Ho and Angela Ryan, "World Scientists' Statement, Update of Concerns," Institute of Science in Society, July, 1999.

• **Is Your Dinner Patented?** Researchers at Johns Hopkins University Medical School reported in 1997 that broccoli sprouts mixed with daikon radish sprouts could reduce induced breast cancer up to 70% in laboratory animals. Following the announcement, the national price of broccoli seeds soared and seeds were in short supply to plant the fall crop. When Sunny Creek Farm in North Carolina began selling broccoli sprouts, it received a registered letter from Brassica Protection Products, warning that the method had been patented by Johns Hopkins University. Brassica, which was formed to enforce the patent, offered to license sprout growing at about 30 cents per container. "To put this into perspective, our wholesale price for alfalfa sprouts is only 40 centers per container," explained Ed Mills of Sunny Creek. "My greatest concern is that this licensing fee will prevent such a good food from reaching the tables of the poor in our nation. I was told that Brassica plans to prosecute anyone who grows broccoli sprouts without their license, including individuals growing in their homes."

Source: "Medicinal Foods or Pharmaceuticals?" *The Scoop*, French Broad Food Co-op, Asheville, NC, August/September, 1998.

• **Landmark Supreme Court Decision:** "Driving the worldwide biotech industry is the patenting of genes and novel life forms and the promise of windfall profits. In 1980 the U.S. Supreme Court upheld an appellate court ruling that 'the fact that microorganisms are alive . . . [is] without legal significance' and ruled that life forms created in a test tube could be patented. This decision, *Diamond vs. Chakrabarty* (later extended by the federal courts to all GM multicellular living organisms, including animals), may be remembered as the most important case of the century—more far-reaching than even *Brown vs. the Board of Education* and *Roe vs. Wade.*"

Source: Alex Jack, "Enlarging the Circle of Life," *One Peaceful World Journal*, Summer 1999.

8. Treatment of People as Intellectual Property

• **U.S. Seeks to Patent Native Blood Factor:** The U.S. has upheld the legality of patenting not only seeds, foods, crops, viruses, bacteria, and other forms of life, but also the genetic material found in human beings. In 1993, researchers from the National Institutes of Health (NIH) discovered that native people from the Guaymi tribe in Panama carried a unique virus in their blood that might prove useful in medical research. Taking a blood sample from a 26-year-old Guaymi woman, the scientists developed a cell line and applied for a patent.

Isidro Acosta, the leader of the native community, said the Guaymi were shocked to learn that their biological inheritance had been stolen by the U.S. government. "I never imagined people would patent plants and animals. It's fundamentally immoral, contrary to the Guaymi view of nature, and our place in it. To patent human material . . . to take human DNA and patent its products . . . that violates the integrity of life itself, and our deepest sense of morality."

Media coverage and public outrage forced the U.S. government to back down. Later when the U.S. tried to do the same thing in the Solomon Islands and Papua New Guinea, South Pacific Island nations joined together and proclaimed their region "a patent-free zone." The American government abandoned the patent effort in 1996.

Source: Jeremy Rifkin, *The Biotech Century* (Tarcher/Putnam, 1998), pp. 57-58.

9. Monopoly Control and the Concentration of Wealth

• **Controlling the World's Seed Supply:** Biotech companies are buying up conventional and organic seed companies around the world in order to monopolize the world's seed supply and eventually replace it with new engineered varieties. In the last several years, for example, Monsanto has acquired or controlled eight major seed companies including Asgrow, Dekalb, Delta & Pine Land, Gargiulo Tomato, Hartz, Holden, Naturemark, and Stoneville Pedigreed.

The companies typically enforce proprietary rights. Hartz, for example, requires farmers to sign the following contract: "The Grower agrees not to supply any of this seed to anyone for planting and agrees not to save any crop produced from this seed for replanting or supply saved seed to anyone for replanting. The grower agrees not to use this seed or provide it to anyone for crop breeding, research or seed production. If a herbicide containing the same active ingredients as Roundup Ultra herbicide (or one with a similar mode of action) is used over the top of Roundup Ready soybeans, the Grower agrees to use only the Roundup branded herbicide."

"In giving the green light to genetic technology, we may be consigning the planet's food production into the hands of a few corporate monopolies whose interest in the bottom line may transcend their commitment to serve the public interest," observes agricultural researcher Marc Lappé.

"By ensuring that seed crops will be selected by the persons who grow them and not by the corporate detail person, we will assure the perpetuation of the deeper linkages that bond the farmer to the soil, and ultimately, the consumer to the natural earth on which we all depend."

Source: Marc Lappé, Ph.D. and Britt Bailey, *Against the Grain* (Common Courage Press, 1998), pp. 53, 150.

• **A Monopoly on Vegetable Seeds:** Alfonso Romo, a Mexican billionaire, has quietly acquired vegetable seed companies over the last five years, cornering the market on lettuce, carrots, tomatoes and other vegetables that are supplied to salad bars across the nation. "Now 40% of all vegetables sold in the U.S. supermarkets derive from Mr. Romo's seeds," the *Wall Street Journal* reported. In an article on monopoly control of seeds, Frederick Kirschenmann, Ph.D., founder of Farm Verified Organic, noted, "The most immediate problem organic farmers face will be the difficulty of obtaining non-transgenic seeds. Even saving seed will be difficult if Monsanto succeeds in pushing new regulations through state legislatures. These would require that seed cleaning operations, which remove weed seeds, broken kernels and other debris, keep samples and records. Seed cleaners would have to test incoming seeds to determine which were bioengineered, enabling Monsanto to verify whether farmers are saving and replanting patented seeds. Many small cleaning plants who handle seed saved by organic farmers may not be able to afford the extra storage and testing facilities. And without access to genetically natural seeds, it will be impossible for certified organic farmers to maintain their natural farming systems, as they are required to use non-transgenic seed."

Source: Frederick Kirschenmann, Ph.D., "Seeds of Crisis for Organic Farms," *The Green Guide*, August 1999.

10. Spread of Reductionist Science and Genetic Determinism

• **The Limits of Modern Science:** Nina Moliver, a nutritionist and macrobiotic teacher in Boston, discussed the science behind genetic engineering at a panel on genetically altered foods at the Kushi Institute 1999 Summer Conference."The underlying assumption of gene technology in food is that technology, as Vananda Shiva puts it, is a superior substitute for nature. Nature is a source of scarcity, and technology is a source of abundance, and the result, as she says, is ecological destruction and new scarcities . . . Technology is no substitute for nature. What this comes down to, is a mechanistic, reductionist view of the world which unfortunately many people in our society understand as science. It is one way of science, but it's not by any means the only way of science or knowing the world.

"This mechanist reductionism understands the world in terms of isolated component parts which don't really interact. So if you have corn and want to make it more nutritious, you add an amino acid to the gene for corn. That makes the corn more nutritious. It lacks context, it lacks holism. What are the other effects of adding this one amino acid to the genome of the corn? What is its context? This kind of scientific thinking uses mechanics—trained mechanics who are good at shooting the gene into the genome—but they don't understand the environment, how health works, or how the body works. . .

"Most established molecular geneticists alive today have a direct or indirect vested interest in a biotechnology enterprise, so that influences their science and it influences the questions that they ask. The answers that you get in science are never any better than the questions you ask. If the question you ask a scientist is how best can I break down the barriers between species, you are going to get one set of answers. If the question is how can I best replenish this topsoil so that it has a lot of earthworms in it and its rich and thick and produces healthy plants on its own, you're going to get a completely different set of answers. . .

Source: Nina Moliver, Kushi Institute Summer Conference, August 1999.

• **Genes Fluid and Dynamic:** "The technology is driven by an outmoded, genetic determinist science that supposes organisms are determined simply by constant, unchanging genes that can be arbitrarily manipulated to serve our needs; whereas scientific findings accumulated over the past 20 years have invalidated every assumption of genetic determinism. The new genetics is compelling us to an ecological, holistic perspective, especially where genes are concerned. The genes are not constant and unchanging, but fluid and dynamic, responding to the physiology of the organism and the external environment, *and require a stable, balanced ecology to maintain stability."*

Source: World Scientists' Statement Calling for a Moratorium on GM Crops and Ban on Patents, Institute of Science in Society.

• **Gene Research a Diversion:** In *Exploding the Gene Myth,* Ruth Hubbard, a Harvard biologist, characterizes the quest to map the human genome as misguided. "By focusing our attention on microorganisms or genes, scientists succeed in drawing attention away from societal influences. They also ensure their own monopoly, by keeping disease prevention in scientific institutes and laboratories. Hubbard, a staunch opponent of genetically altered foods, says in her book that emphasis on genes detracts from promising solutions such as diet and lifestyle.

Source: Ruth Hubbard and Elijah Wald, *Exploding the Gene Myth* (Beacon Press, 1993).

• **Genetic Engineering Corrupts Science:** Refuting the modern dogma that genes are fixed and determine all life processes, R. C. Lewontin a geneticist at Harvard, says that biotechnology has corrupted science, diverted attention from underlying causes of sickness, such as diet, and resulted in a massive waste of public funds such as the genome project. "It has been clear since the first discoveries in molecular biology that 'genetic engineering,' the creation to order of genetically altered organisms, has an immense possibility for producing private profit. . . . No prominent molecular biologist of my acquaintance is without a financial state in the biotechnology business."

Source: R. C. Lewontin, *Biology as Ideology: The Doctrine of DNA* (Harper, 1992).

11. Increased Tolerance of Ethnic Cleansing

• **"Undesirable" Genes Compared to "Undesirable" People:** In an essay touching on the war in Kosovo and Serbia, Alex Jack, a macrobiotic teacher at the Kushi Institute and director of the One Peaceful World Society, observed, "On a genetic level, modern science is attempting to perform what political and military leaders are carrying out on a geopolitical scale. The campaign to eliminate undesirable viruses, bacteria, genes, and other microscopic phenomena parallels the way in which nation-states are trying to 'cleanse' entire ethnic groups, religions, and tribes. We can avert this crisis if we awaken to the awareness that diversity, whether in the environment, society, or DNA, is positive and strengthens and enriches the whole.

"Secondly, brute force, whether on a battlefield, in the home, or in the laboratory, is counterproductive. Ethnic cleansing and genetic cleansing must stop. Respecting life in all its manifestations leads in the long run to balance, stability, health, and peace. Rather than demonizing those who differ from us—whether statesmen or terrorists, welfare mothers or alienated teens, 'harmful' microbes or 'undesirable' genes—we should reflect and see that they are contributing to overall harmony and balance, usually as a result of our own excesses.

"When you exclude something or someone from a gene pool, family, school group, or the community of nations, it easily turns into an outcast or demon and fulfills your worst expectations. Far better to include everyone in the family of humanity, the commonwealth of nations, and the circle of life. By striving to harmonize with nature and improving our own limitations, together we can arrive at a peaceful, comprehensive solution."

Source: Alex Jack, "Enlarging the Circle of Life," *One Peaceful World Journal*, Summer 1999.

12. Increased Aggression and Violence, Especially Among Children

• **Are GM Foods Contributing to School Violence?:** "In the last couple of years, violence in the schools has become as American as apple pie. In fact, the dramatically altered quality of hamburgers, french fries, apple pie, and other staples in the modern diet may be at the root of the new wave of violence, especially among children," observed Alex Jack, a health care teacher at the Kushi Institute and co-author of *The Cancer-Prevention Diet*.

"Pesticides, for example, have been linked by researchers with a variety of symptoms, including irritability, aggression, and violence. Since most genetically altered foods contain Bt, a built-in pesticide, their consumption may be contributing to the epidemic of violence.

"Indeed, a majority of foods on American supermarket shelves—and school cafeterias—now include GM ingredients and hence an increased burden of pesticide. The FDA, whose job it is to ensure the safety of American food, does not regulate GM foods because it does not have the authority to regulate pesticides. The EPA, whose job is to regulate pesticides, won't require a warning label on Bt-laced corn, soybeans, or potatoes because it has no authority over foods. Normally, the EPA requires a bottle of Bt to carry a label warning people not to inhale the substance and to avoid getting it in an open wound.

"Troubling as this Alice-in-Wonderland mentality is, there are also energetic effects of consuming new modified foods, which also may be contributing to the widespread outbreak of violence, especially with guns.

"Genetically altered organisms are originally created by placing the desired gene in a solution and muzzle-loading it into a small-caliber pistol. These 'microbullets' coated with DNA are then fired into a screen covering targeted plant cells and tissue in a petri dish, penetrating the cellulose and dispersing the DNA into the cells.

"The method of firing DNA into cells was developed at the University of California in Davis and originally involved shooting regular .22 or .45 mm bullets. 'The gun is mounted in a box and the barrel is extended so that the bullet travels a couple of feet and then hits a metal block that has a slit in that is smaller than the bullet,' explains Catherine Willet, who researched the first experiments and now teaches biology at MIT. 'The bullet is stopped by the metal block, but the little plastic balls are not. They are small enough to fit through the slit, so they keep traveling at the original speed of the bullet through the slit. The plant cells are mounted on the other side of the slit, so they are bombarded by the little plastic balls which can penetrate the cellulose and carry the DNA into the cells, where a tiny fraction of it is incorporated into the plant cell nuclear and transforms the cell. The transformed cells are then grown into a plant; the seeds from this plant are then the source from which genetically altered plant strains are obtained.'

"From the traditional and holistic view of 'you are what you eat,' we absorb the energy and vibration of the foods we take in, as well as the nutrients and other material components. Macrobiotic cooking, for example, emphasizes keeping a calm, peaceful mind in every step of the agricultural and food preparation process. High quality food is not only pure—whole, natural, and chemical-free—but also it is grown, processed, and prepared in a harmonious spirit.

"Traditionally farmers serenaded their crops, people used slow, natural methods of processing, and cooks put strong, healthful energy into their cooking with their thoughts and prayers. Today, of course, the opposite prevails. Modern food is produced, refined, and processed through high-energy methods that artificially speed up natural processes of fermentation, ripening, and curing. In the kitchen, microwave ovens, electric ranges, and other high-energy devices create a chaotic vibration. Homes are often battlegrounds between family members, cooking is avoided as much as possible, and anger, abuse, and neglect often spill over into the kitchen. All of these negative influences are directly absorbed by the food and passed along to all who eat it.

"Genetically altered food is especially weak and devitalized because it lacks the strong, health-giving charge that has evolved naturally for millions of years. The fact that it is born literally in the muzzle of a gun, rather than in the bosom of the Earth, should give us pause. It is disturbing to think what energetic effect consuming foods originating from gun-fired DNA will have on human cells and tissues, organs and functions, and minds and consciousnesses, especially those of young people. Brave New World meets Chairman Mao ('Power comes from the barrel of a gun') in the high-tech supermarket and school cafeteria.

"It would be an enlightening scientific experiment to examine the foods served in the Littleton, Colorado high school and elsewhere where violence has broken out. These menus and recipes could then be compared with those in school districts where violence is not a problem and with schools that feature natural selections.

"This spiral of violence will end when parents, teachers, and communities awaken to the destructive nature of the foods they are feeding our children. Berkeley's switch to organic foods is a role model for the nation and a beacon for a healthier, more peaceful world."

Sources: Alex Jack, "Dietary Roots of Violence," *One Peaceful World Journal*, Autumn 199, p. 39; Catherine Willet, email, August 6, 1999.

• **Berkeley School District Goes Organic and Dumps GM:** Berkeley, California approved a plan in the summer of 1999 that would convert its school cafeterias to all-organic food and ban BGH, genetically altered foods, and irradiated foods. "This is part of a whole effort to improve nutrition and educate our kids and our families about good nutrition," explained Karen Sarlo, a spokeswoman for the Berkeley Unified School District.

The organic plan would add pesticide-free vegetables and fruits as well as whole grain sandwiches to the menu. Students will grow some of their own food in school gardens and local organic farmers will supply the rest. "The whole philosophy is to have food become part of the education," said Jered Lawson, who is training teachers how to incorporate lessons on food and gardening into literature, science, and math classes.

Source: "From Berkeley, Calif., A New School Innovation," Reuters, August 18, 1999.

• **Gene-Fired Food Unstable:** The process of firing DNA into plants is inherently unstable. Referring to "the shotgun approach" of genetic engineers, agricultural researcher Marc Lappé notes, "This scattershot approach has its downside: No one knows where the novel DNA has been spliced into the plant's own array of genes. Only by blink luck will a few plants survive, and those that do will have this new DNA in different regions of their genome."

Dr. Arpad Pusztai, a British geneticist, also emphasized how imprecise and unreliable the method is, "Think of William Tell shooting an arrow at a target. Now put a blindfold on the man doing the shooting and that's the reality of the genetic engineer when he's doing a gene insertion. He has no idea where the transgene will land in the recipient genome."

Sources: Marc Lappé, Ph.D. and Britt Bailey, *Against the Grain* (Common Courage Press, 1998), pp. 27, 30; and "Why I Cannot Remain Silent: Dr. Pusztai Talks to *GM-Free*," *GM-Free* 1:3, August/September 1999.

13. Violent Orientation toward Women, the Feminine, and Nature

• **Rape of Nature:** Nina Moliver, a nutritionist in Boston, stresses the sexual distinction between genetic engineering and traditional cross-breeding. "To say it is like traditional cross-breeding is like saying that going out on a date with someone is to have someone violate you. Or inviting guests into your home is the same thing as having a robber force his way into your home and take your belongings. These foreign genes that could not possibly combine in nature are shot in with a gun—violently. It is an act of violence. It supersedes the natural process of sexual reproduction, so it's anti-sexual in its nature. It removes nature from the process, it removes environmental selection from the process. It's not possible to make this multibillion dollar investment pay for itself unless people come back and buy these seeds year after year from the same company. So saving seeds has to become illegal or made impossible by various means. The whole process of the interaction of the seed with its environment is lost or ruined. When you take two batches of identical seeds and put them in two totally different ecosystems, in five years you will have two slightly different varieties, because each one will have interacted with its own ecosystem. So there's no interaction left. The seed comes from the laboratory only. The environment has nothing to do with it. Sexual reproduction is eliminated."

Source: Nina Moliver, Kushi Institute Summer Conference, August 1999.

• **"Genetic Engineering" a Euphemism:** "The entire concept of 'genetic engineering,' 'genetic modification,' and 'genetic alteration' is a euphemism, concealing the inherent violence of the process. These terms are essential neutral and value-free, while the reality is that scientists are forcing unwanted genes into virgin nature. I would think that everyone, but especially women, would be appalled at the brute force inherent in this process, as nature is commonly regarded as feminine. Calling GE, GM, and GA 'genetic evisceration,' 'genetic manipulation,' or 'genetic abduction' would be a more accurate description. Certainly Dante would have found a special niche for this practice in his moral universe. Terms such as 'biopiracy' and 'bioserfdom' convey some of the magnitude of this onslaught.

Source: Alex Jack, Kushi Institute Summer Conference, Panel on Genetically Altered Foods, August 1999 and other lectures.

14. Abuse and Alteration of Language

• **Liberty® or Death for American Farmers:** In an article comparing national reporting of the war in Kosovo and the dangers of genetically modified foods, author and teacher Alex Jack observed, "In 'ethnic cleansing,' 'collateral damage,' and other euphemisms that conceal terror and killing, Orwellian doublespeak is alive and well. . . The current issue of *Science News* contains a revealing statement in the wake of last month's chilling research report that Monarch butterflies are endangered by genetically altered corn: 'Biotech companies sell corn carrying the toxin gene, designed to protect the crop from mother caterpillars *with minimal collateral damage* to bees and other beneficial insects.'"

"Monsanto has been granted patents on their Bt gene technology for use in Bollguard® cotton, Yieldgard® corn, and New Leaf® potato products, effectively disguising the real nature of these products and their effects on health and the environment," he observed in another essay. "But probably the most outrageous example is glufosinate, the pesticide used in Bt corn and other modified products and their companion herbicides. It is sold to farmers under the brand name Liberty™. Some of the patents for altered Bt, incidentally, are held by Ecogen, whose name, like that of many 'life science' companies, conveys respect for the natural environment.

"The misuse and manipulation of language is mirrored on the other side of the debate as well. Environmental superheroes who take on the establishment—portrayed as mega polluters, brokers of endangered species, and other evil cartels and conspiracies committed to world domination—are emerging in fiction, TV, film, and other media, creating a simplistic, comic book world view. In San Francisco, for example, a young woman calling herself Super Gene Girl disrupted a Monsanto-sponsored conference. Stripping off her biohazard suit and declaring that she would rather go naked than wear GM cotton, she shouted, 'Gene-spliced cotton is not sustainable!'

"Militant anti-GA activists who destroy fields of engineered crops style themselves as 'Eco-Warriors' or 'Croppers' and wrap themselves in the mantle of patriotism and environmentalism. There is an admirable Boston Tea Party like aura to these provocative, relatively nonviolent actions. But as we experienced in the peace movement during the Vietnam era, direct action has a way of escalating into violent destruction of scientific laboratories and government facilities. Inevitably, if it continues, some people will be injured or killed. Thus David changes into Goliath!

"The natural integrity of all structures and institutions—beginning with the images and language of consciousness and extending to personal health, the family, the community, and the world as a whole—has steadily weakened over the last century. About fifty years ago, the modern supermarket began to do away with all natural environmental boundaries by making foods from around the world available regardless of climate or season. Today, as the 21st century begins, modern fusion cooking is violating ecological boundaries, bringing together exotic foods and beverages that have never been combined before. New digital technologies now allow us to completely alter text, photographs, graphics, and sounds and take them out of their original context. The introduction and consumption of genetically modified foods is contributing to the new cut-and-paste mentality in society that threatens to end millions of years of natural evolution, rewrite the language of our own genes, and lead to unprecedented planetary chaos."

Source: Alex Jack, "Enlarging the Circle of Life," *One Peaceful World Journal*, Summer 1999 and Alex Jack, "The Alteration and Splicing of Language," September, 1999.

• **The Vocabulary of Pollution:** In *The Ecology of Eden*, social critic Evan Eisenberg describes how language is being misused in the present environmental debate. "At the far end are the crazed futurists, the sort of people who would make cows legless milk-dispensers, conveniently stackable, and replace sheep with tube-fed lamb-chop cultures hundreds of yards long. . . . Also at the far end are such recent movements as 'Wise use' and 'takings,' which use the language of freedom and property rights to disguise sheer greed and stupidity. For these people, science is just window dressing. Virtually the only scientists on their side are the lab-coated mannikins they pay to stand behind the glass. . . . Between the gaily colored advertisements in which major polluters express their commitment to a cleaner planet can be found advice on the management of water, air, population, energy, industry, economic development, agriculture, biodiversity, and climate. . . ."

Source: Evan Eisenberg, *The Ecology of Eden* (New York: Alfred Knopf, 1998), p. 431.

15. Development of Biological Weapons and Increased Risk of Bioterrorism and Biowarfare

• **Genetic Threat Replaces Nuclear Peril:** The introduction of genetically engineered seeds, crops, and foods contributes to theoretical and technological advances that could be employed in the creation of biological weapons. Some scientists and futurists are predicting that genetic warfare may replace nuclear warfare as the greatest military threat to global security. For example, crops can just as easily be engineered to produce blight, hunger, and disease as their opposite.

"The new genetic engineering technologies provide a versatile form of weaponry that can be used for a wide variety of military purposes, ranging from terrorism and counterinsurgency operations to large-scale warfare aimed at entire populations," explains social critic Jeremy Rifkin. "Unlike nuclear technologies, genetically engineered organisms can be cheaply developed and produced, require far less scientific expertise, and can be effectively employed in many diverse settings."

Source: Jeremy Rifkin, *The Biotech Century* (Tarcher/Putnam, 1998), pp. 2, 93.

• **British Medical Association Warns of Misuse of the Human Genome for Military Purposes:** Several months before coming out publicly against genetically altered foods, the British Medical Association addressed a related issue, warning that biological and genetic weapons would pose a serious threat in the near future. At a press conference warning of the dangers of misuse of biotechnology, Dr. Vivienne Nathanson, the head of health policy research at the BMA, said, "It would be a tragedy if in 10 years time the world faces the reality of genetically engineered and possibly genetically targeted weapons." She explained that "designer" weapons are under development that exploit genetic variations to target its victims." For example, microbes could be created in a laboratory to attack known receptor sites on the membrane of cells or viruses could be targeted at specific DNA sequences inside cells. "An ethnically targeted weapon becomes more of a reality," she said. At the press conference, BMA officials launched a new book, *Biotechnology Weapons and Humanity*, encouraging strengthening of the 1973 Biological and Toxin Weapons Convention and calling on physicians and scientists to carefully monitor the mapping of the human genome and protect the integrity of their work.

Source: "Biological 'Ethnic' Weapons Loom on the Horizon, Reuters, January 21, 1999.

6
Ethical, Religious, and Spiritual Hazards of Genetically Modified Foods

"Love is not love
which alters when it alteration finds . . . "
—Shakespeare, SONNET 116

1. Playing God and Regarding Humans as Creators of Life

• **Religious Leaders Form Coalition:** A national coalition of several hundred American religious leaders expressed their opposition to the granting of patents for genetically engineered life forms in 1995. The group included the leaders of most major Protestant churches, more than 100 Catholic bishops, and Jewish, Muslim, Buddhist, and Hindu officials. Organized by The Foundation on Economic Trends, the coalition said that the patenting of life marked the most serious challenge to the notion of God's Creation in history. The religious leaders strongly disagreed with classifying life as a human invention and allowing scientists and corporations to profit from it.
Source: Jeremy Rifkin, *The Biotech Century* (Tarcher/Putnam, 1998), p. 65

• **Anglicans Deny Church Lands to GM Trials:** The Church of England refused to allow the government of the U.K. to test genetically modified crops on its lands. After meeting with Agriculture Ministry officials, the Church Commissioners, who manage the church's property, denied permission to set up experimental field trials pending a full-scale inquiry into "the theological implications" of genetic engineering.
Source: "Church of England Bars GM Crop Trials," Reuters, August 4, 1999.

2. Violation of Dietary Codes

• **Religious Lawsuit Filed against FDA:** Representatives of major religions in the U.S. filed a lawsuit in 1998 against the FDA charging that the lack of labeling of GM foods makes it impossible for practitioners to observe religious dietary codes and customs. The plaintiffs, including several Jewish rabbis, a Catholic priest, a Seventh-Day Adventist minister, an Eastern Orthodox cleric, and a Buddhist, demanded mandatory testing and labeling of GM foods and products. According to Steven Druker, a lawyer who is party to the suit, in the Bible Leviticus 19:19 forbids the mating of one species of animal with another, as well as sowing a field with two types of seeds. Mixing different species in food results in "destroying the natural boundaries of natures," adds Rabbi Jossi Serebryanski, a Hasidic rabbi in Brooklyn.

"Viewed from the Eastern Orthodox theological perspective, this process appears to be utilizing the infectious, destructive forces of nature in the creation of new life forms, which seems like a gross affront to the Creator's original design," stated Father Samuel Kedala, an Eastern Orthodox priest at the Holy Spirit Orthodox church in Wantage, N.J.

Rev. DeWitt Williams, director of health ministries for the Seventh-day Adventist church in North America, expressed concern because half of the church's 10 million members are vegetarians and exposed to GM soybeans.

Altogether, parties to the lawsuit included 113 Christians, 37 Jews, 12 Buddhists, and 122 people who said their "faith is not easily categorized."

Source: Robert S. Greenberger, "Religious Groups Push for FDA Labels on Biofoods," *Wall Street Journal*, August 17, 1999.

3. Danger of Genetic Apocalypse

• **A Molecular Auschwitz:** Genetic engineering is "a molecular Auschwitz," according to Dr. Erwin Chargoff, biochemist and father of molecular biology. "I have the feeling that science has transgressed a barrier that should have remained inviolate. . . . You cannot recall a new form of life . . . It will survive you and your children and your children's children. An irreversible attack on the biosphere is something so unheard of, so unthinkable to previous generations, that I could only wish that mine had not been guilty of it."

Source: Dr. Erwin Chargoff, *Heraclitean Fire* (Rockefeller University Press, 1978).

4. Violation of Natural Order

• **Violating Species Integrity:** In a panel on genetically altered food, Edward Esko, a macrobiotic teacher and author of *Healing Planet Earth,* focused on the violation of natural order. "Each species has its own niche, its own environment, its own food. Each species is distinct from all other species. We don't see in nature species crossing over. We don't see elephants mating with giraffes. Strange, isn't it? We don't see crocodiles mating with chimpanzees. Species have an integrity. If you talk to any person, even a child, that is common sense, species integrity. It is as common sense as the sun coming up every morning. But because of our very narrow, reductionist, scientific view, backed by huge commercial interests, we have as a species, as human beings, violated our own most basic common sense.

"With genetic engineering, we have crossed a Rubicon. We have violated the most fundamental law and order of nature which is species integrity. Not only species integrity, but kingdom integrity. By inserting fish genes into a tomato, for example, we have blurred the distinction between the plant and animal world. This is a recipe not for the order of the universe, the order of nature, but chaos and disorder and destruction. This is the end of nature, and on the spiritual level, this is the height of arrogance. As every moral and historical lesson teaches us, the more arrogant and self-important we become, the more we set ourselves up for a huge fall.

"Once a reporter went up to the North Shore area of Boston after a huge hurricane came in and did millions of dollars of damage. He filed his report and at the end said, "No matter how you look at it, nature always bats last." Nature is always the home team, it gets the last chance to even the score. It's common sense. We can't oppose, go against, or ultimately change this order of the universe. It will bat last and will even the score, even though it looks like our position, our group, our grassroots effort is very small. We don't have the billion-dollar budgets, but we do have the power of truth. We have the fact that everybody on the earth today, regardless of whether healthy or sick, regardless of vegetarian or meat-eater, everyone on the earth today is still the product of natural genes and species' integrity. Even if they are eating beef, that beef still has integrity. But that is disappearing. The fact that everyone of us comes from that background means that somewhere in the recesses of everyone's conscience, there is that glimmer of awareness and that basic common sense. That comes out every time there is a debate about labeling these products. Labeling is a great fear of the biotech industry, because they know that most people still have common sense and would tend to reject their foods. And most people, surveys say, will insist on labeling. However, if these processes continue and more people eat altered foods, that reservoir of common sense will start to be lost. We will lose our chance to turn this planet around. The critical time is now, before this becomes the deluge. This is something like the biological flood of Noah spreading all over the earth. Before this is unleashed fully, we have another 15, 20, or 30 years to awaken and enlighten everyone and redirect human resources and human inventiveness.

"Once, in the Chicago airport, I met a young lady who turned out to work for a top genetic research company in London. We had a very spirited discussion over a Guiness. I started to explain our view and asked her, "What about your colleagues? Is there any questioning about this?" She said, "Not at all. They love it. They think this is the cutting edge. This is where the real creative lines of the late 20th century are converging. They are totally into it." I tried to show her, point by point, the back side to it. But I came away from that conversation realizing that we can't just simply try to stop genetic engineering and deny that power and the attraction of human beings to that creative process. We need to rechannel that, to create something equally challenging, in a really natural way, to use that intellectual power. Always throughout history when intellectuals have carried on from that level, without the thought of social consequences, it has been a recipe for disaster. In this century, the most notable example was the atomic bomb. Those creative scientists were working for that, and they unleashed nuclear power, a hugely destructive force. Really what we need in a democracy is an enlightened population. We can't depend on elected politicians. Everybody has to elevate their consciousness, elevate their awareness, and make their voices heard."

Source: Edward Esko, Kushi Institute Summer Conference, August, 1999.

7
Scientists, Physicians, and Environmentalists Speak Out

"If we possessed a thorough knowledge of all the parts of the seed of any animal (e.g. man), we could from that alone, by reasons entirely mathematical and certain, deduce the whole conformation and figure of each of its members, and, conversely if we knew several peculiarities of this conformation, we could from those deduce the nature of its seed."

—Descartes, OEUVRES IV

1. On Genetic Engineering

• **What Is Genetic Engineering?:** "DNA contains a complete set of information determining the structure and function of a living organism, be it a bacterium, a plant, or a human being.

"DNA is a very long string of 'code words,' arranged in an orderly sequence. It contains the instructions for creating all the proteins in the body.

There is no way to make a gene insert in a predetermined location. So the insertion is completely haphazard.

"In mating . . . the sequence of DNA 'code words' in each chromosome remains unchanged. And the chromosomes remain stable. The mating mechanism has been developed over billions of years and yields stable and reliable results.

"In genetic engineering, a set of foreign genes is inserted haphazardly in the midst of the sequence of DNA 'code words.' The insertion disrupts the ordinary command code sequence in the DNA. The disruption may disturb the functioning of the cell in unpredictable and potentially hazardous ways. The insertion may make the chromosomes unstable in an unpredictable way.

"So technically, genetic engineering is an unnatural insertion of a foreign sequence of genetic codes in the midst of the orderly sequence of genetic codes of the recipient, developed through millions of years. This is a profound intervention with unpredictable consequences."

Source: "What Is Genetic Engineering?" Physicians and Scientists for the Responsible Application of Science and Technology (PSRAST).

• **Nobel Scientist Speaks Out:** "Up to now, living organisms have evolved very slowly, and new forms have had plenty of time to settle in.

Now whole proteins will be transposed overnight into wholly new associations, with consequences no one can foretell, either for the host organism, or their neighbors. . . . going ahead in this direction may be not only unwise, but dangerous. Potentially, it could breed new animal and plant diseases, new sources of cancer, novel epidemics."

Source: Dr. George Wald, Nobel Laureate in Medicine, 1967 and Professor of Biology, Harvard University.

• **Altering the Genetic Code:** "A handful of people have decided to change the genetic code of the world's plant and animal kingdoms without knowing the consequences, and they haven't bothered to ask permission from the rest of us."

Source: Robert D. Klauber, physicist, quoted by Mothers for Natural Law.

• **Rewriting the Genetic Library:** "Soy used in infant formula has genetic material and bacteria never before ingested by the human race. We are rewriting the genetic library of the earth in only three to five years."

Source: Dr. John Hagelin, director of the Institute of Science, Technology, & Public Policy, quoted by Mothers for Natural Law.

2. British Medical Association Calls for Moratorium on GM Foods and Crops

• **Multiple Health Effects Cited:** In 1999, the British Medical Association called for a moratorium on GM crops and foods in Britain and a ban on the import of unlabeled foods from abroad. The report, entitled "The Impact of Genetic Modification on Agriculture, Food, and Health," recommended that GM crops should not be released into the environment until they are deemed safe. The reported called for a ban on the use of antibiotic resistant marker genes in GM food, segregation of regular crops from GM products to prevent cross-pollination, and mandatory labeling.

"Antibiotic resistance, the threat of new allergic reactions, and the unknown hazards of transgenic DNA mean that on health grounds alone the impact of genetically modified organisms must be fully assessed before they are released," a BMA spokesperson noted. "The environmental implications and therefore the long-term effects on human health cannot be safely predicted at this stage and caution must therefore prevail."

Commenting on the report, Sir William Asscher, Chairman of the BMA's Board of Science and Education, stated: "Once the GM genie is out of the bottle, the impact on the environment is likely to be irreversible. That is why the precautionary principle is so particularly important on this issue. It is even more serious than the licensing of medicines, which can, if necessary, be withdrawn. That is why the BMA is pressing for an open-ended moratorium until there is much greater scientific certainty about the risks and potential benefits of GMOs."

Dr. Vivienne Nathanson, BMA Head of Health Policy and Research, added: "Antibiotic resistance, the threat of new allergic reactions and the unknown hazards of transgenic DNA mean that on health grounds alone the impact of GMOs must be fully assessed before they are released. The environmental implications and therefore the long term effects on human health cannot be safely predicted at this stage and caution must therefore prevail."

The BMA has 119,000 members and represents more than 80 percent of the physicians in the U.K.

Source: British Medical Association, May 18, 1999.

3. Institute of Science and Society Calls for a Moratorium on GM

• **Statement on Life and Evolution:** "Life is an intimate web or relations that evolves in its own right, interfacing and integrating its myriad of diverse elements. The complexity and interdependence of all forms of life have the consequences that the process of evolution cannot be controlled, though it can be influenced. It involves an unpredictable creative unfolding that calls for sensitive participation from all the players, particularly from the youngest, most recent arrivals, human beings.

"Life must not be treated as a commodity that can be owned, in whole or in part, by anyone, including those who wish to manipulate it in order to design new life forms for human convenience and profit. There should be no patents on organisms or their parts. We must also recognize the potential dangers of genetic engineering to health and biodiversity, and the ethical problems it poses for our responsibilities to life. We propose a moratorium on commercial releases of genetically engineered products and a comprehensive public enquiry into the legitimate and safe uses of genetic engineering. This enquiry should take account of the precautionary principle as a criterion of sensitive participation in living processes. Species should be respected for their intrinsic natures and valued for their unique qualities, on which the whole intricate network of life depends.

"We recognize the validity of the different ways of knowing that have been developed in different cultures, and the equivalent value of the knowledge gained within this traditions. These add substantially to the set of alternative technologies that can be used for the sustainable use of natural resources that will allow us to preserve the diversity of species and to pass the precious gift of life in all its beauty and creativity to our children and their children, to the next century and beyond."

Source: "Life and Evolution," statement signed by scientists associated with the Institute of Science in Society (ISIS).

4. Physicians and Scientists for the Responsible Application of Science and Technology Warn About GM

• **Gigantic Bay of Pigs Syndrome:** "It is time for the U.S. government to swallow the bitter truth—that the world is realizing that the launch of genetically engineered products on to the market is highly premature and unjustifiable. It is time for the U.S.—as well as the industry—to take global responsibility for the environmental and health safety of its products and apply a precautionary approach to the introduction of new technologies.

"The world is already suffering from the damage from products launched before sufficient knowledge has been gathered about their safety. Genetic engineering is incomparably the most profound manipulation of nature ever done. As released genes cannot be recalled, harmful consequences are irreparable and may spread indefinitely. Very little is known about the environmental consequences. Yet the U.S. uses all its power to force the world to accept is worldwide application blindly. It is unrealistic to believe that this will be successful at length. Already investors are becoming increasingly hesitant, for good reasons . . .

"The U.S. GE food project appears more and more as a gigantic "[Bay of Pigs] syndrome," where the government is pressing towards an inevitable catastrophe assisted by biased industry-dependent 'scientific' advisors."

Source: "Genetically Engineered Foods—Safety Problems," Physicians and Scientists for Responsible Application of Science and Technology (PSRAST), August 16, 1999.

5. Sierra Club Letter to President Clinton on GEOs and the Environment

• **Conservation Group Warns of Environmental Threat:** In an open letter to President Clinton, the Sierra Club, the largest grassroots conservation group in the United States, expressed concern with the safety of genetically engineered organisms. "Our purpose is to protect the ecosystem and we believe that the rate of application of this technology far exceeds our ability to understand the environmental and public health risks and to avoid potentially serious impacts," Laurel Hopwood, chair of the Biotechnology Task Force, wrote.

The letter disputed claims by the biotech industry that genetic engineering is a simple extension of traditional crossbreeding techniques. "While conventional breeders face natural barriers that prevent unrestricted gene transfer between unrelated species, genetic engineers bypass this protective barrier by combining genes from totally unrelated species. Furthermore, the technology involved in transferring foreign genes is imprecise, unstable, and unpredictable, so that engineers have no way of predicting how GEOs will behave once released into the environment."

The Sierra Club went on to call for:

• Extensive, rigorous research on the potential long term environmental and health impacts of GEOs before they are released into the environment

• Use of the precautionary principle, whereby: 1) harm is avoided before scientific certainty has been established, and 2) the burden of proof is shifted to those with the power and resources to prevent harm

• Mandatory environmental impact statements to be made for every ecosystem into which any new GEO is to be introduced. These should be based on rigorous science and open public debate.

• An end to the concept of "substantial equivalence" by our regulatory agencies as a ploy to sidestep safety studies and oversight responsibilities. For example, toxins meant to kill insects are being genetically engineered into plants, yet the consequences of these toxins in the diets of humans, livestock, beneficial insects, and wildlife are unknown.

• Mandatory labeling of genetically modified products after full safety assessment is completed and doing so in a manner that is easily discernible. All consumers, both citizen and corporate, should be given the right to choose what they buy.

• Removal of all antibiotic resistance genes from all food crops, which are routinely placed in genetically engineered crops. It is recognized that such extensive use of antibiotic market genes is unnecessary and will likely hasten the development of antibiotic resistant pathogens, depriving us of one of the most profound accomplishments of 20th century medicine.

• U.S. commitment not to use trade negotiations or agreements to override the rights of countries to regulate GEOs. The launch of new talks on biotechnology at the upcoming Seattle Summit of the World Trade Organization should not take place without thorough, open, and participatory environmental assessments conducted parallel to the negotiations.

• Full U.S. ratification of the Convention on Biological Diversity, already ratified by 175 other nations, and forceful leadership to support its goal of protecting the diversity of life on Earth. Recognition that biodiversity is not a luxury but a foundation of life on our planet.

The statement further cited the threat to Monarch butterflies, and noted that "pollinator species, such as bees, may themselves be harmed, with disastrous consequences to the food supply."

In conclusion, the letter stated, "The Sierra Club calls for the expansion of research into the risks that recombinant DNA technology and its products pose to the natural environment. In the meantime, in the absence of scientific knowledge, the Sierra Club asks that we take a precautionary approach. Until rigorous research is conducted to discern and address the long term impacts of GEOs, particularly in regards to their use in agriculture, such organisms should not be released into the environment."

Source: Sierra Club Letter to President Clinton, August 18, 1999.

6. Testing Issues and Concerns

• **Impossibility of Testing:** Richard Lacey, a medical doctor, microbiologists, and Professor of Food Safety at Leeds University, became world famous after predicting the made cow epidemic in the early 1990s. He has strongly opposed GM foods and crops because of "the essentially unlimited health risks." "The fact is, it is virtually impossible to even conceive of a testing procedure to assess the health effects of genetically engineered foods when introduced into the food chain, nor is there any valid nutritional or public interest reason for their introduction."

Source: quoted by Mothers for Natural Law.

• **Animal Testing Model May Be Faulty:** Professor Dennis Parke of the

University of Surrey School of Biological Sciences and a British advisor to the U.S. FDA on the safety aspects of biotechnology has called for a moratorium on the release of GM organisms, foods, and medicines. "In 1983 hundreds of people in Spain died after consuming adulterated rapeseed (canola) oil. This adulterated rapeseed oil was not toxic to rats."

Source: quoted by Mothers for Natural Law.

• **No Way to Track GM Consumption:** Without labeling, it will be difficult, if not impossible, for epidemiologists to determine the effects on human populations. "Should there be a biological effect from ingesting altered soybean by-products (as may result from changes in their estrogenic components), we presently lack any way to identify or track affected populations," noted health researcher Marc Lappé. He also expressed concern whether "new patterns of pesticide use would create health risks to farm workers or consumers?"

Source: Marc Lappé, Ph.D. and Britt Bailey, *Against the Grain* (Common Courage Press, 1998), pp. 2-3.

7. FDA Challenged

• **FDA Scientists Question GM:** While the FDA officially holds that there is no evidence of any harmful effects of GM foods, documents obtained under the Freedom of Information Act by the Alliance for Bio-Integrity, a community activist organization, show otherwise.

Steve Druker, a lawyer and spokesman for the Alliance, noted, "In internal documents FDA officials repeatedly cautioned that foods produced through recombinant DNA technology entail different risks than do their conventionally produced counterparts and that this input was consistently disregarded by the bureaucrats who crafted the agency's current policy, which treats bioengineered foods the same as natural ones. Besides contradicting the FDA's claim that its policy is science-based, this evidence shows the agency violated the U.S. Food, Drug and Cosmetic Act in allowing genetically engineered foods to be marketed without testing on the premise that they are generally recognized as safe by qualified experts."

"There is a profound difference between the types of unexpected effects from traditional breeding and genetic engineering which is just glanced over in this document," the documents quoted Dr. Luis Priybl of the FDA Microbiology Group. Several aspects of gene splicing ". . . may be more hazardous."

According to the newly released data, Dr. Linda Kahl, an FDA compliance officer, objected that the agency was "trying to fit a square peg into a round hole . . . [by] trying to force an ultimate conclusion that there is no difference between foods modified by genetic engineering and foods modified by traditional breeding practices. The processes of genetic engineering and traditional breeding are different, and according to the technical experts in the agency, they lead to different risks."

Dr. Jim Maryanski, an FDA Biotechnology Coordinator, acknowledged that there is no consensus about the safety of GM foods in the scientific community and FDA scientists advised they should undergo special testing, including toxicological tests.

Source: "FDA Documents Show They Ignored GMO Safety Warnings from Their Own Scientists," Alliance for Bio-Integrity, June 24, 1999.

8. Fallacy of Substantial Equivalence

• **FAO/WHO Standard Challenged:** In 1996, the Food and Agricultural Organization (FAO) and World Health Organization (WHO) introduced the concept of substantial equivalence, which has been widely applied to approving genetically engineered foods. "Substantial equivalence embodies the concept that if a new food or food component is found to be substantially equivalent to an existing food or food component, it can be treated in the same manner with respect to safety (i.e., the food or food component can be concluded to be as safe as the conventional food or food component."

In an assessment of the concept, scientists Mae-Wan Ho and R. Steinbrecher found that the principle is:

• intentionally vague and ill-defined to be as flexible, malleable, and open to interpretation as possible

• comparisons are designed to conceal significant changes resulting from genetic modifications

• the principle is weak and misleading even when it does not apply, effectively giving producers carte blanche

• there is insufficiency of background information for assessing substantial equivalence

• there is no specification of tests for establishing substantial equivalence

• there is no requirement to test for unintended effects and that current tests are undiscerning and may even serve to conceal unintended effects

"We accept that no safety assessment system is foolproof," the researchers concluded. "A case in point is the rigorous testing that goes on with pharmacological products. It is estimated that despite such rigorous testing, 3% of the products approved for market turned out to have such harmful effects that they have to be withdrawn, while an additional 10% have sufficiently harmful side effects that limited use has to be recommended. This underlines the importance of segregation, clear labelling, and postmarket monitoring of the health and other impacts of genetic engineered foods. Labeling is a matter of traceability, especially for the case of potential allergenicity, and should be a scientific requirement, not only a consumer option."

Source: M.W. Ho and R. Steinbrecher, "Fatal Flaws in Food Safety Assessment: Critique of the Joint FAP/WHO Biotechnologhy and Food Safety Report," *Environmental & Nutritional Internations* 2:51-84, 1998.

• **Fallacy of Substantial Equivalence:** In an interview, Dr. Arpad Pusztai, a noted British geneticist, observed: "The idea of 'substantial equivalence' is that there is no need for biological safety tests because the plants must be of similar composition as the parent line. This is the basis on which GM crops are being released. However, they cannot be substantially equivalent to the parent because you've introduced new genes."

Describing an experiment in which he compared two transgenic lines of potato produced from the same gene insertion and the same growing conditions, he continued, "We found that one of the lines contained 20% less protein than the other. So the two lines were not substantially equivalent to each other. But we also found that these two lines were not substantially equivalent to their parent. This could not be predicted. It demonstrates that the unpredictability is inherent in the GM process on a case by case basis and also at the level of every single GM plant created."

Source: "Why I Cannot Remain Silent," *GM-Free* 1:2,August/September 1999.

9. Flawed Testing of Roundup Ready

• **Flawed Testing:** In a review of the testing procedure for Roundup Ready soybeans, Judy Carman, Ph.D., MPH., an epidemiologist and senior lecturer at the Research Centre for Injury Studies, Flinders University, in Southern Australia, found that the feeding trials performed by Monsanto were deeply flawed.

The pesticide Roudup works by inhibiting an enzyme needed by the plant to synthesize certain aromatic amino acids, thus killing the plant. The enzyme, known as EPSPS, is incorporated into a bacterial version of this enzyme into the soybean plant so that spraying with Roundup will kill the weeds but not the soybeans.

To test for the effects of digesting this modified protein in humans, Monsanto set up an in vitro (artificial) mamalian gastric and intestinal mixture but did not test it on living subjects. "The documents also states that as people cook soybeans before consumption, this would deactivate the enzyme and thus people would not consume it," Carman notes. "However, raw soybeans will be fed to cattle. Steak is often served medium rare to rare. Therefore, there is a possibility that people will consume this new still-functional enzyme in their diet." She observed that Monsanto did not measure quantities of this enzyme in cattle tissue, the ability of the enzyme to persist during moderate cooking, or the effect it would have on animals as well as humans. Nor was lecithin, a common soy product used as an emulsifier in food, tested in the assessment. Moreover, the soybeans used in the study were not those harvested from farms but laboratory specimens.

Further laboratory on rats were also flawed, according to Dr. Carman. Not only was the sample small and statistically insignificant, but "they appear to have done no biochemistry, immunology, full autopsy, histology (except on pancreas), etc." In respect to immunological effects of the soybeans, she further noted, no human trails were conducted to test their safety.

Citing the need for independent studies, Dr. Carman called for chemical analyses of Roundup Ready soybeans from farmers' fields, long-term feeding studies, randomised, double-blind feeding trials with human volunteers, long-term cohort studies on consumers in countries where these foods are already been sold and eaten.

"In summary," she stated, "I believe that the scientific basis, provided by the applicant company for considering that Roundup Ready soybeans are safe for animal and human consumption, is seriously flawed. No other, independent investigations seem to have been done. It could be expected that the safety assessments of other genetically modified foods may be as flawed.

"Independent testing of these foods is urgently required, incorporating long-term animal and human experiments. As they will take years, it would be wise to place on a moratorium on these foods for 5 years, as suggested by European groups, while these investigations are done. To do otherwise could be likened to permitting a giant feeding experiment on millions of people."

Source: Judy Carman, Ph.D., M.PH, "The Problem with the Safety of Roundup Ready Soybeans," Flinders University, Australia, 1999.

• **Flawed Testing:** Commenting on the Roundup Ready soybean tests, Dr. Arpad Pusztai, the British geneticist, stated, "GM foods have been introduced on the back of just one published paper. Just one, in 15 years of GM. It was written by a Monsanto scientist and published in 1996. The study was a feeding trial of Roundup Ready soya on rats, catfish, chicken, and cows. I don't want to say anything about it because it's a published paper, but I could take it apart in 10 seconds."

He went on to say, for example, that Monsanto used mature animals which are not forming body tissues and organs. "Adults need only a small amount of protein because their bodies are in equilibrium, in homeostasis. But a young growing animal needs a great deal more protein because it's laying down muscle and tissues, and forming its organs. With a nutritional study on mature animals, you would never see any difference in organ weights even if the food turned out to be anti-nutritional. The animals would have to be emaciated or poisoned to show anything."

He also took issue with the post-mortems. "They never weighed the organs; they just looked at them—what they call 'eyeballing.' I must have done thousands of post-mortems so I know that even if there is a difference in organ weights of as much as 25%, you wouldn't see it."

Source: "Why I Cannot Remain Silent," *GM-Free* 1:2, August/September 1999.

10. Field Testing Unethical and Potentially Harmful

The British government has established experimental field trails of herbicide-tolerant genetically altered corn and rapeseed on several farms in the UK. Dr. Mae-Wan Ho, a scientist who has warned of the danger of horizonal gene transfer from GM crops to unrelated species, opposes the tests as unethical. "It is irresponsible for the Government to continue with the massive farm-scale field trials in view of the evidence its own scientists are taking into account," she noted. "Transgenic pollen can travel for several miles, she went on to say. It will affect farm workers, food processors, and expose the general public to transgenic DNA, while bees will certainly take it up and contaminate the honey." The current farm-scale trials carry no provision to monitor for horizontal gene transfer or impacts on health, she charged. The

GM corn, moreover, carries a "disrupted" ampicillin-resistance gene, which is not expressed, Ho added. Given the mutability of the altered gene, she said that it may become reactivated in bacteria.

Source: "MAFF Reveals New Scientific Findings Confirming Fears Over Health Hazards of GMOs," Institute of Science in Society, July 27, 1999.

11. Scientific Hubris

• **Basic DNA Research Never Done:** No scientific data exists documenting the stability of any line of genetically engineered organisms or in the structure and location of the genes inserted in the genome. In principle, regulatory authorities and the biotech industry would provide information on the levels of genes expressed, a genetic map and DNA base sequence of the insertion, and the site of the insertion in the host genome in each successive generation. No such basic research has ever been done.

Source: Mae-Wan Ho, "Dangerous Liason—Deadly Gample," *Agricultural Biotechnology and Environmental Quality: Gene Escape and Pest Resistance*, Biotechnology Council Report #10, Ithaca, N.Y., pp. 105-20, 1999.

• **A Titanic Folly:** "Because the Titanic was declared to be unsinkable, the risks from icebergs were thought to be negligible. Scientists today are equally compartmentalized in their thinking."

Source: Jean-Marie Pelt, director, European Institute of Ecology, Green Conference on Genetic Engineering, Brussells, Belgium, March 9, 1998.

• **Looking Backward (If There Is a Future to Look Back From):** "Future 'experts' who look at our crash programs for allowing one herbicide like glyphosate to dominate agriculture in the late 1990s will be astounded at our hubris and ignorance. There is little doubt that this herbicide has fewer visible side effects and safety problems in terms of acute toxicity than many others. But its saturation of the environment is almost certainly going to have adverse ecological effects, especially at the microecosystem level, effects with which a future generation will have to cope."

Source: Marc Lappé, Ph.D. and Britt Bailey, *Against the Grain* (Common Courage Press, 1998), pp. 2-3.

• **Vegetarian Bride of Frankenstein:** In the second volume of his series on the creators of the modern mind, Alex Jack writes about how Mary Shelley came to write *Frankenstein*, the world's greatest horror story and prophecy about the dangers of modern science. Combing visual diagnosis, Nine Star Ki, and the I Ching, he explores why Mary and her husband, poet Percy Shelley, adopted a vegetarian way of eating, and the role diet plays in her famous novel. "In today's age of cloning and genetically-altered food, *Frankenstein, or the Modern Prometheus* (its fully title) constitutes a grim prophecy of what lies ahead if humanity overreaches and, stealing fire from the gods, embarks on a quest for artificial nourishment and love."

Source: Alex Jack, *Profiles in Oriental Diagnosis II: Vegetarian Bride of Frankenstein* (One Peaceful World Press, 1998), p. 98.

8
Political Initiatives and Community Actions

"Europe is so important to the industry that it could mean we'll really have to pull back on growing GM crops in this country. Given the choice, who wants to grow GM?"
—Walt Fehr, head of Iowa State University's biotech department, September, 1999

1. U.S. and Allies Oppose Treaty Controlling Genetically Altered Products

• The world's first international treaty to regulate genetically modified foods and products collapsed when the United States and five agricultural exporting nations rejected a proposal that had the support of the rest of the 130 nations participating in the conference in Cartagena, Colombia. The Biosafety Protocol would have required advance approval from the importing nation before GM plants, seeds, or other organisms were introduced.

"It's five nations against the world," said Dr. Joseph M. Gopo, a Zimbabwe delegate, referring to America and which allies. "There could be no greater injustice than that." The U.S., he asserted, "is holding the world at ransom." The other nations to oppose the treaty were Canada, Australia, Chile, Argentina, and Uruguay. All the European, African, and Asian countries supported the initiative.

Source: Andrew Pollack, "U.S. and Allies Block Treaty on Genetically Altered Goods," *New York Times*, February 25, 1999.

2. Legislation or Lawsuits to Control or Limit Genetically Modified Crops and Foods

• **North America:** In 1998, Maine became the first state to prohibit the use of a GM product—corn incorporating an herbicide-resistant gene from Bt.

• In spring 1999, the New Hampshire Senate voted to ban the use or sale of "terminator" seed technology, seeds, or crops in the state.

• In 1999, the Foundation on Economic Trends founded by Jeremy Rifkin, the National Family Farm Coalition, and farmers from across North and South America brought an antitrust suit against Monsanto, DuPont, and Novartis for trying to gain control of world agriculture.

• In 1999, Canada prohibited cows from being injected with Bovine Growth Hormone (BGH). In autumn 1999, Canada announced voluntary standards for GM food.

• **Western Europe:** In 1994, the European Union banned BGH.

• In 1998, the EU passed a law requiring that all GM food must be labeled.

• In 1998, Austria and Luxembourg banned altered Bt corn. France banned the planting of altered corn but permitted its import if labelled.

• By 1998, 1200 schools in Britain took GM foods off their menus.

• In 1999 European Union environmental ministers in effect implemented a three-year moratorium on the approval of any new GM foods or crops. "We've had a de facto moratorium, and now it's been cast in stone," EU Commission spokesman Peter Jorgenson stated.

• **Eastern Europe:** In 1999, Ukraine declared that it would not accept GM crops, because it did not want any more Chernobyl-like accidents.

• **Asia:** India's Supreme Court ruled in 1999 that all field trials of GM cotton must be halted. The ruling came after months of protests by farmers and consumers who organized a Monsanto Quit India movement styled along Gandhi's famous Quit India movement earlier in the century. The government also banned imports of BGH.

• Japan suspended the approval of Bt crops for agricultural purposes in 1999, following the report that Monarch butterflies are at risk from GM corn. In the first half of 1999, Japan imported 2.11 tons of soybeans from America, accounting for 86% of its total usage. Under consumer pressure, Japan approved mandatory labelling of 22 varieties of six GM crops and foods starting in 2001, including tofu, miso, and most soy products.

• South Korea enacted legislation in 1999 requiring labelling of GM foods.

• In Thailand, the government intends to introduce GM cotton, papayas, tomatoes, chillies, and vegetables and import altered U.S. soybeans.

• In China, an estimated 1 million acres of altered cotton was planted beginning in 1998. "We prefer not to discuss the issue," a Chinese Foreign Ministry official told Reuters. Another said China's policy was a "state secret." (Reuters, September 1, 1999)

• **Africa:** Environmentalists have taken the lead in opposing Terminator Technology and in supporting the Biosafety Protocol.

• **South America:** A Brazilian court in 1999 stopped GM releases in Brazil for one year. Earlier agriculture ministers in all 27 Brazilian states unanimously sent a statement to their federal counterpart asking him not to allow GM crops in Brazil. Opposition is growing in Chile, Paraguay, and Uruguay.

• **South Pacific:** In 1999 Australia and New Zealand decided on mandatory labelling of all GM foods and is conducting tests on 20 products. In New Zealand, Eli-Lilly was forced under strong public pressure to withdraw an application to distribute Monsanto's BGH.

• **United Nations:** In 1999 the GMTT Codex Alimentarius, a UN body regulating international food standards, declined to certify BGH as safe.

• Several UN members, including India, Norway, Ecuador, and the Ivory Coast, have sought to establish a moratorium on biotechnology through the world organization.

3. Breaking Ranks: Multinationals, Food Companies, and Farmers Dump GM

• In Britain, most major food manufacturers, distributors, and sellers have repudiated GM foods and ingredients. By early 1999, even McDonald's, Burger King, and Kentucky Fried Chicken responded to customer pressure by eliminating GM soy and corn from their menus.

• Unilever and Nestle, two of the world's largest food companies, agreed in 1999 to phase out sales in the U.K. of products made with GM ingredients.

• Cadbury, Sainsbury, Safeway in the U.K., Carrefour in France, Pryca in Spain, and Migros in Italy and other major food retailers pledged to eliminate GM ingredients from their brands.

• Gerber Foods, the largest baby food manufacturer in the U.S. and a division of Novartis G. A. in Switzerland, pledged in the summer of 1999 to being phasing out GM soy and corn ingredients and substitute organically grown products. H. J. Heinz, Co., the maker of Earth's Best line, followed suit, while Healthy Times Natural Foods has switched from canola oil, which is commonly modified, to safflower oil.

• Archer Daniels Midland, Co., a giant U.S. commodities processor and merchandiser, agreed to separate conventionally grown corn and soy from GM crops for export to Europe. A. E. Stanley Manufacturing Company, another large corn producer, also began separating its products. ADM, known as "the supermarket to the world," announced in 1999 that it would pay farmers premium prices for non-GA soybeans.

• In Europe, Iams, one of the world's largest pet food companies, rejected the seven varieties of altered corn not approved by regulators. This closes off the major remaining outlet for sales of American-grown GM corn in Europe.

• Honda Trading Corp. announced plans to build a plant in Ohio in autumn 1999 to sort and bag GM-free soybeans. Thesoybeans will be exported largely to Japan for sale to tofumakers. Honda, a subsidiary of Hondo Motor Co., contracted with 118 farmers to supply an estimated 15,000 tons of the soybeans.

• Kirin Brewing Company in Japan announced that starting in 2001 it would use only non-GA corn. The following day, Sapporo, its main competitor, followed suit.

• In Mexico, Grupo Maseca, the leading producer of corn flour for tortillas, announced that it would avoid importing GM corn from the U.S. Total sales of corn to Mexico amount to $500 million annually.

• In South Korea, a large importer of American grain indicated that it was considering purchasing non-GA products from China instead.

• In England, retreating from GM, AstraZeneca announced in autumn 1999 that it might sell its agrichemicals division. In Switzerland, Novartis, the world's largest maker of crop products, did likewise.

4. Labeling

• **Microbiologist Urges Caution:** Dr. John Fagan, a microbiologist and cancer researcher at Maharishi University of Management, was one of the first scientists to advocates a precautionary approach and call for mandatory labeling of all GM foods. "Without labeling, it will be very difficult for scientists to trace the source of new illness caused by genetically engineered food," stated Fagan, who turned down a $3 million government research grant to call attention to the dangers of biotechnology.

Source: Mothers for Natural Law.

• **Prince Charles Speaks Out Against GM and for Labeling:** "The public discussion so far has concentrated on the risks and capabilities of the technology and the effectiveness of the regulations. These things are important, as are effective and comprehensive labelling schemes to ensure that those consumers like me who do not want to eat GM foods can avoid them."

Source: The Prince of Wales, St. Jame's Palace, December 1998.

• **U.S. Secretary of Agriculture Speaks Out on Labeling:** A staunch supporter of genetic engineering, Dan Glickman, the Secretary of Agriculture, gave a conciliatory speech in which he raised the issue of voluntary labeling for the first time and the need for the biotech industry to be more candid about its products. In his speech and follow up comments to the press, he was reported to have stated:

"What we cannot do is take consumers for granted . . . a sort of 'if you grow it, they will come' mentality."

"We can't force-feed consumers . . . There are certainly more and more questions being asked about biotechnology, and those questions must be answered."

"Some type of informational labeling is likely to happen."

Source: Dan Glickman, National Press Conference, Washington, D.C., July 13, 1999.

• **Consumers Union Calls for Labeling:** "The U.S. requires labeling orange juice 'from concentrate' and vegetables as 'frozen,'" said Jean Halloran, director of the Consumer Policy Institute, a division of Consumers Union. "Ignoring 'genetically engineered' threatens to undermine public trust in a labeling system millions rely on every day."

Source: "Consumer Group Urges Labeling of Modified Foods," Reuters, August 24, 1999.

• **Congressional Call to Action:** Dennis J. Kucinich, a Congressman from Cleveland, Ohio, spoke out strongly against GM foods and introduced the first legislation to study the health and environmental effects as well as require labeling. After French farmer protested against McDonald's restaurants for using altered ingredients, Kucinich told *Time*, "The farmers in France are right. There's nothing more personal than food." "When butterflies start dying," Kucinich, one of the few members of Congress eating macrobiotically, added, "I think it's fair to start asking questions."

Source: Jeffrey Kluger, "Food Fight: The Battle Heats Up Between the U.S. and Europe over Genetically Engineered Crops," *Time*, September 13, 1999.

5. Boycotts, Direct Action, and Guerrilla Theatre

• **GM Crops Destroyed in Europe:** In Europe, direct action against genetic engineering spread across the continent in the late 1990s. In Germany, squatters occupied several fields to prevent the planting of GM sugar beets in Freiburg. In France, farmers sabotaged and destroyed a consignment of Novartis's GM maize seed in Nerac. In Britain, protesters dumped 4 tons of GM soybeans on the doorstep of 10 Downing Street, the residence of Prime Minister Tony Blair, a staunch advocate of biotechnology. A costumed "Frankenstein" often appears outside grocery stores warning that "Franken Foods" are sold inside. In Ireland, activists from Genetic Concern uprooted a GM crop of sugar beet under cultivation by Monsanto in Teagasc, Carlow, and pledged to strike elsewhere.

• **Indian Fields Set Ablaze:** In South India, 200 farmers set two test plots of genetically altered crops on fire under development by Monsanto.

• **California Experiment Uprooted:** At the University of California in Berkeley, the California Croppers pulled up a field of GM corn being tested by Novartis, the biotechnology giant, on Thanksgiving, 1998, and again in July, 1999. The crops were under the supervision of the Plant Gene Expression Center, a collaboration of the USDA and the University of California, Berkeley. Meanwhile, in San Francisco, activists calling themselves fabRAGE (Fabulous Resistance Against Genetic Engineering) stripped off their clothes made of genetically engineered cotton and disrupted a panel featuring a spokesman from Monsanto.

• **Seeds of Resistance:** A half acre of GM corn was destroyed in Old Town, Maine, by environmental activists in August, 1999. The field was part of an experiment conducted by the University of Maine. An estimated 1000 stalks of corn were destroyed with a machete. On the Internet, the Seeds of Resistance, a group associated with Earth First!, later claimed responsibility for the action.

• **Nature Strikes Back:** In England, a fox ruined an experiment on genetically altered crops by raiding insect traps filled with alcohol and eating the beetles and bugs that had fallen into them. The experiment was set up to show how engineered crops can encourage wildlife and reduce pesticide use among farmers. "This young fox obviously liked drinking the alcohol, eating the pickled insects, and getting fairly drunk in the process," Dr. Alan Dewar, an entomologist associated with the experiment, explained.

Source: "GMO-Drunken Fox Threat to GM Crop Study," *Telegraph*, U.K., August 5, 1999.

• **Mutant Cow Dung:** Residents of Craig Country, Va. opposed Pharming Healthcare Inc's plan to introduce a herd of 200 cows and produce human proteins in transgenic milk because they feared genetically-altered manure could pollute their groundwater.

Source: "Viriginia Residents Oppose Transgenic-Cattle Farm," *Richmond Times-Dispatch*, September 22, 1999.

9
What Can I Do?

"To offset the rise of cloning, genetic engineering, and AIDS, Ebola, mad cow disease, and other new epidemics, humanity may need a stronger staple food to resist the effects of bionization and artificial genetic influences."

—Michio Kushi and Alex Jack, HUMANITY AT THE CROSSROADS

1. Dietary and Lifestyle Guidelines

1. Eat carefully, selecting foods that are not genetically modified. Eat whole foods, naturally and organically grown, as much as possible.These include whole grains, beans, vegetables, sea vegetables, fresh fruits, seeds, nuts, and other, predominantly plant-quality foods. Several natural foods growers, manufacturers, or distributors are now labeling their foods and products as free of genetically modified organisms and ingredients. These include Eden Foods, Hain Food Group, and Ben & Jerry's.

2. Start a garden and use natural seed, especially traditional heirloom seeds. These are also known as open-pollinated or standard seeds and are available through selected natural seed companies. Save seeds and share with friends (which also will cut down on "genetic drift" to your garden!)

3. Make your own tofu, tempeh, miso, bread, baked goods, salads, pasta and noodle dishes, and other naturally processed or prepared foods as much as possible from whole, natural foods. When you purchase these foods ready-made, as much as possible seek out naturally or traditionally processed foods made with natural or organic ingredients.

4. Use high-quality organic vegetable-quality cooking oil such as unrefined sesame or corn oil. Avoid canola oil and products with canola oil.

5. Avoid commercially grown corn, potatoes, tomatoes, and yellow squash when shopping or eating out. Read labels carefully and avoid non-organic products that include corn syrup, corn starch, corn oil, corn fructose, corn dextrose, soy oil, soy flour, soy protein, lecithin, and other products containing corn or soy.

6. Avoid milk, dairy, meat, poultry, and other livestock products as much as possible, especially any non-organic products. Most dairy cows, beef cattle, pigs, sheep, and chickens are fattened on animal feed that includes GM ingredients or organisms.

7. Avoid commercially grown cotton, which may be genetically modified, as much as possible. Use organic cotton or other natural fabrics instead.

2. Community Action

1. Ask your local supermarket, restaurant, co-op, and other outlets not to carry genetically modified foods. As a first step, request that they survey their suppliers on whether they use GM ingredients and make that information available to consumers for an informed choice.

2. Talk to local parents and school officials and request that your local school district initiate a policy of avoiding GM foods in its school lunch program. In Berkeley, California, the school district voted to serve only organic foods free of GM and irradiated products after local parents, teachers, and environmentalists mobilized the community.

3. Contact your representatives at the local, state, and federal levels and ask, at a minimum, for mandatory labeling and ideally for a moratorium on GM crops and foods pending comprehensive testing on their long-time health and environmental effects. Responding to public concern, the New Hampshire Senate recently voted to ban Terminator Technology from being sold, used, or grown in the state.

4. If you have investments, consider divesting your portfolio of stocks in biotech companies or use your stocks to go to stockholder meetings and bring up issues related to health and the environment.

5. Keep informed about issues related to genetic engineering, through books, periodicals, and the Internet (see Resource chapter).

3. Meditation and Self-Reflection

1. Meditate and pray for universal health, happiness, and peace. Reflect on your own diet, environment, and way of life and whether they are contributing to a world of peace and harmony. Make positive changes.

2. Visualize the world awakening to the dangers of genetic engineering and taking strong individual and collective action to return to organic and sustainable agriculture. Visualize Monarch butterflies and other insects, animals, and plants endangered by biotechnology living freely. Imagine a world without Monsanto, or a Monsanto peacefully transformed into an organic company (see Appendix 1 "Imagine a World Without Mansanto").

3. Walk outdoors and especially absorb the natural energy of ripening whole grains, vegetables, and other crops.

4. Sing a happy song each day and listen to harmonious, calming music, especially Gregorian chant, Hildegard of Bingen, shakuhachi flute, and other traditional music; Bach, Mozart, and other Baroque and Classical composers; spirituals and hymns; and any other music that moves you.

5. Avoid extreme reactions to the enormity of the GM issue, including hopelessness, apathy, and fear of every morsel you eat. Find a middle way. Be mindful of what you eat without being fanatical or fatalistic.

6. Trust in nature and 4 billion years of natural evolution to help humanity through this unprecedented crisis.

7. Keep a calm, peaceful mind and stay balanced.

4. Guidelines and Home Cares to Counter the Effects of Genetically Modified Foods

• **Organic Seeds Build DNA:** To rebuild DNA that is weakened by GM foods, macrobiotic educator Michio Kushi recommends eating organically grown grains and seeds. The most important are whole grains such as brown rice, millet, and barley. Next are seeds, especially sesame seeds and pumpkin seeds. These will help restore DNA. The seeds can be prepared in various ways. Other macrobiotic-quality healing foods are also recommended, especially mineral rich sea vegetables such as kombu, wakame, nori, arame, and hiziki, regularly; miso soup daily made with organic GM-free soybeans; and ume-sho-kuzu tea occasionally (but not daily).

Source: Michio Kushi and Alex Jack, *Humanity at the Crossroads: Dietary and Lifestyle Guidelines forthe Age of Cloning, Genetically Engineered Food, Mad Cow Disease, Microwave Cooking, Electromagnetic Fields, & Global Warming* (One Peaceful World Press, 1997), pp. 99-100.

5. Proposals to Support Biodiversity

• **Ecological Sustainability:** In a review of the environmental effects of GM foods, Miguel A. Altieri, a professor in the department of environmental science, policy and management at the University of California, stated that engineered crops and foods pose a serious threat to agro-ecosystems. To protect the environment, he recommended:

• Public funding of research on transgenic crops that enhance agrochemical use and that pose environmental risks be ended

• HRCs and other transgenic crops should be regulated as pesticides

• All transgenic food crops should be labeled as such

• Funding for alternative agricultural technologies should be increased

• Ecological sustainability, alternative low-input technologies, the needs of small farmers and human health and nutrition should be pursued with greater vigor than biotechnology

• Measures should be taken to encourage sustainable and multiple use of biodiversity at the community level, with emphasis on technologies that promote self-reliance and local control of economic resources as a means to foster a more equitable distribution of benefits

Source: Miguel A. Altieri, "The Environmental Risks of Transgenic Crops: An Agroecological Assessment," *Pesticides and You*, Spring/Summer 1998.

6. Cultivating Ancient, Heirloom, Open-Pollinated, and Other Traditional Seeds

• **Seeds of a New Civilization:** "For humanity to survive from generation to generation, its DNA must remain strong and healthy," educators Michio Kushi and Alex Jack write in *Humanity at the Crossroads*. "Despite the rise of the health revolution and spread of organic foods, seed quality is steadily declining. Most organic foods are grown today with hybrid seeds. Natural foods are gradually losing their ki, or natural electromagnetic energy, because of the intensifying environmental crisis, chemical contamination, and artificial seeds. As seed quality declines, the functioning of the brain, intestines, and ultimately the reproductive organs, including the sperm and eggs, are affected. It is essential to secure the best quality seeds and food. The seeds must be cultivatable." The introduction of genetically altered seeds, foods, and products takes the decline of the modern food supply to a new, unprecedented level.

"Many people are asking whether ordinary organic crops will be able to meet the biological challenge. We are hopeful that our macrobiotic community, along with the natural foods and organic agriculture movements, will be strong enough in the years ahead to offset the evolutionary crisis. Let us hope that traditional heirloom seeds that have nourished humanity for thousands of generations can return to being the standard and that organic farmers, food processors, natural food stores, and natural foods consumers will resist the temptation to use less expensive hybrid seeds."

Recently some rice seeds dating back thousands of years have been discovered in ancient tombs in Japan. The rice comes in many colors, including black, red, brown, and yellow, similar to traditional Indian corn. It gives very strong energy, and among the different types the black is the strongest. The rice comes from northern Japan and dates back to the Jomon era, an ancient culture known for its large burial mounds and the world's earliest weaving. "In addition to the high energy of the planet at that time which was absorbed by the seeds," the authors explain, "the burial mounds have served to charge the seeds for thousands of years. Moreover, the ancient rice is dry land rice. That means it does not need water irrigation, just a little moisture to grow. It is more like wild grain that can be scattered and will grow almost anywhere. It's more strengthening than paddy field rice.

"For usual good health and recovery from many chronic conditions, such as cancer and heart disease, regular brown rice is more than adequate. However, to offset the rise of cloning, genetic engineering, and AIDS, Ebola, mad cow disease, and other new epidemics, humanity may need a stronger staple food to resist the effects of bionization and artificial genetic influences."

The ancient rice is available in limited quantities in Japan and is being experimentally grown by farmers in the United States. One Peaceful World and the Kushi Institute hope to make the ancient rice freely available in the next several years.

Source: Michio Kushi and Alex Jack, "Seeds of a New Civilization," *Humanity at the Crossroads: Dietary and Lifestyle Guidelines for the Age of Cloning, Genetically Engineered Food, Mad Cow Disease, Microwave Cooking, Electromagnetic Fields, & Global Warming* (One Peaceful World Press, 1997), pp. 104-105.

- **Sources of Traditional, Non-GM Seeds:**

Heirloom Seeds, Box 245, W. Elizabeth PA 15088, (412) 384-0852. www.heirloomseeds.com

The Kusa Society, Box 761, Ojai CA 93024. Enclose $2.50 SASA for info.

Seed Savers Exchange, (319) 382-5990.

Seeds of Change, Box 15700, Santa Fe NM 87506; (888) 762-7333.

Southern Exposure Seed Exchange, Box 170, Earlysville VA 22936, (804) 973-4703.

Appendices

1. Imagine a World without Monsanto

MEMO TO MONSANTO
To: Robert B. Shapiro, CEO Monsanto Corporation
From: Alex Jack, CEO One Peaceful World
Re: Merger

Dear Mr. Shapiro,

I was fascinated to read a profile of you in today's *New York Times* business section. Your nontraditional approach to corporate culture, including wearing sweaters, khakis, and casual shoes to work and networking with employees, is refreshing. I was especially impressed with your hobby of playing Go, the Japanese board game based on yin/yang strategy.

The reason I'm e-mailing you is that I understand Monsanto is seeking a merger. Over the last several years, Monsanto has become the world's leading producer of genetically engineered seeds, foods, and products (known colloquially as Frankenfoods). Evidently, research and development—not to mention environmental lawsuits—do not come cheaply. Last year, amid mounting losses, your company agreed to be bought by American Home Products for $34 billion plus change, but the deal collapsed.

To get to the point, I propose we merge our two organizations. OPW is the world's leading international macrobiotic information network. We have publishing, educational, Internet, and travel divisions, as well as corporate ties in agriculture and food production.

What's in it for you? Materially, not much. My office in the Stone House on the Kushi property takes up about 75 square feet, so that wouldn't give your company much room to expand from its headquarters in St. Louis which includes 26 greenhouses and 176 labs.

As for other tangible assets, I have a LaserWriter II printer on my desk that I've used since 1988 (before terminator technology invaded the computer industry), and a hanging scroll of Kwan Yin, the Chinese Goddess of Mercy. She's not such a bleeding heart Immortal as her epithet suggests. For stealing a handful of peaches from heaven, she imprisoned the Monkey King under the Mountain of the Five Elements. (What do you think the cosmic karma is for stealing or cloning entire peach fields?)

Actually, OPW's greatest assets are visionary. We have an imperishable dream of returning to a more natural way of life and creating a world of enduring health and peace. Macrobiotic ideas have spread around the world over the last generation, enhancing thousands, possibly millions, of lives. Attached is an e-mail file containing the text of my new book, *Let Food Be Thy Medicine*, which summarizes the major scientific and medical studies on macrobiotics and holistic health over the last thirty years. (I'll send you a hard copy, if you prefer.) In addition to helping prevent cancer and heart disease—chronic disorders that kill about 75 percent of people—a macrobiotic diet has been found to help reduce aggressive and antisocial behavior, increase fertility, and benefit the environment. Dr. T. Colin Campbell, CEO of the China Health Study; Dr. William Castelli, CEO of the Framingham Heart Study; and other nutritionists support our way of eating. The Smithsonian is opening a permanent collection on macrobiotics and alternative medicine this spring in recognition of its impact on society.

Our product, so to speak, is *macro bios*—life itself—which, contrary to conventional corporate wisdom, can't be marketed, patented, or engineered. Our silent partner, the O.U. (Order of the Universe) is absolute, invincible, and, when challenged, merciless. Four billion years of natural evolution will not be reversed by some funny-looking guys (or gals) in white coats (or sweaters) monkeying around with test tubes. So despite our anemic financial ledger, we are going with the natural order, which is to respect the diversity of the Earth and cultivate its endless natural abundance.

The beauty of the proposed merger is that our side, the macrobiotic community, will handle product development and quality control, while your side takes care of business. Frankly, Bob (if I may call you that), like most holistic communities, macrobiotics sucks when it comes to making money.

For us, relieving PMS, unhardening arteries, and preventing tumors is child's play compared to running a successful company. With Monsanto's 32,000 employees, $28 billion in stock, and annual revenues of $9 billion, you are in a strategic global position to market the natural diet and health revolution. The end result will be to transform Monsanto/One Peaceful World (MOPW) into a real "life sciences" company.

Of course, our deal would mean an end to biotechnology and replacing genetically modified products with the best quality organic and natural seeds and foods. Your addiction to creating and patenting new life forms is simply bad business. That attempt to introduce GM cotton in the South resulting in thousands of acres of deformed crops and millions in damages was a fiasco. Then there was the recall of canola seeds containing an unapproved gene planted in 600,000 acres in Canada. But that's all toxic water under the bridge.

The truth is Monsanto needs a makeover. Despite the laid-back office environment (eating too many modified corn chips lately?), your corporate image is stuck somewhere between the tobacco companies and nuclear power. I'm sorry but the hundreds of Dr. Frankensteins you employ will simply have to be downsized following our corporate marriage. However, we will retrain them to become macrobiotic cooks, chefs, and shiatsu practitioners. (Please check our references with Horst Schulze, CEO of the Ritz-Carlton, which is now serving macrobiotic food in its hotels worldwide.)

FYI, our chairman, Michio Kushi, recently brought back some ancient rice seeds found in a tomb in Japan. These seeds give very strong energy and may be able to naturally offset the wave of new viral epidemics, prion diseases (such as mad cow), and other new scourges that are sweeping the planet—all without adding Bt to their genes.

If Monsanto were to develop and distribute these seeds, it would do wonders for your top and bottom lines. Instead of being reviled as the company that created PCBs, Agent Orange, BGH, Ready Round Up soybeans, Bt-laced corn, and Terminator Seeds, you would be universally cheered. Greenpeace would welcome tanker ships with your new natural products into Europe with an escort of whales and dolphins. African farmers would sing your praises, and the Bio-Safety treaty would be ratified. To paraphrase Monsanto's GM slogan, "Let the *Natural* Harvest Begin."

Meanwhile, with our sister organization, the Kushi Institute, we can arrange to give you exclusive rights to administer Kushi macrobiotic seminars, cooking classes, health consultations, and other educational activities around the world and in cyberspace. Together, we could help realize humanity's age-old dream of a world without war, disease, or hunger. Untold generations would benefit as natural evolution continues to unfold.

By the by, Bob, the photo in the *Times* (nifty vest!) shows a slight bulge and redness on your chin. In traditional Oriental medicine and philosophy, this area corresponds with the prostate and developing disorders in that region. You really ought to consider coming to the Way to Health program at the Kushi Institute ASAP.

Despite our different approaches to sowing and reaping, from the view of the O.U. we are actually going in the same direction. Please consider this proposal—a yin/yang match made in heaven—seriously. To be honest, my intuition tells me that Monsanto will probably merge instead with DuPont, Dow Chemical, or some mega polluter, and I will probably go back to my antique printer and churn out books and manifestos on the perils of biotechnology. A golden moment for our two organizations will be lost.

Be careful, however, as you go about expanding the genetic marketplace and staking out new life science territories. Nature is not easily deceived. As an experienced Go player, you know very well that just when you think you have encircled the opponent, you discover—too late—that you are encircled instead.

Source: Alex Jack, "Memo to Monsanto," *One Peaceful World Journal*, Spring 1999.

2. Open Letter to Dan Glickman, Secretary U.S. Department of Agriculture

The Smithsonian Institution recently honored macrobiotics for pioneering the natural, organic foods movement and alternative medicine in the U.S. A permanent collection of books and literature, natural foods, cookware, and other items was opened in ceremonies at the National Museum of American History on June 9, 1999. The display will be followed by regular symposiums, conferences, research grants, and other activities related to the continuing impact of diet on shaping America's health and consciousness.

As American citizens and individuals and families trying to eat in a healthier direction, we are deeply concerned with the quality of food in this country and would like to thank you for your recent decision to authorize the first independent, long-term scientific studies on the effects of genetically modified (GM) foods. We are also grateful for your support of voluntary labeling of these foods and for your support of national organic standards, including the decision not to allow GM foods to be labeled organic.

We request that these positive steps be quickly followed up by mandatory labeling of GM foods to allow consumers to make a fully informed decision on what they choose to eat and feed their families. We also respectfully call upon you to issue a moratorium on the further use of GM seeds, crops, foods, or products, pending the results of comprehensive long-term studies on human health and the environment. Future generations will honor you for taking such a prudent, far-sighted stand.

—Signed by more than 100 members and supporters
of the One Peaceful World Society, Summer 1999

Resources

One Peaceful World

• **One Peaceful World** is an international information network directed by Alex Jack. Member benefits include a free book from OPW Press, a subscription to the quarterly *OPW Journal* with news of GM foods, organic standards, and macrobiotic and vegetarian recipes, and book discounts. $30.00/year. OPW, Box 10, Becket MA 01223, (413) 623-2322, fax (413) 623-6042.

Daily News on GM -Related Issues

• **Reuters**, the British news agency, consistently has the best coverage of GM events in the U.S. as well as Europe, Asia, and worldwide. On the Web, it can be accessed through the Yahoo GM Food Debate News Page: www.yahoo.com/Full_Coverage/Science/Genetically_Modified_Food/

Essential Books

• Kushi, Michio and Alex Jack, *Humanity at the Crossroads* (One Peaceful World Press, 1997). Dietary and lifestyle guidelines for genetically altered foods, mad cow disease and other new epidemics, electromagnetic fields, and global warming.

• Rifkin, Jeremy, *The Biotech Century* (Tarcher/Putnam, 1998). Excellent overview of genetic engineering and its impact on modern society, culture, and civilization.

• Lappé, Mark and Britt Bailey, *Against the Grain* (Common Courage Press, 1998). An outstanding investigative study of biotech agriculture by a veteran public health researcher and his associate.

• Ho, Mae-Wan, *Genetic Engineering: Dream or Nightmare* (Gateway Books, 1998). An impassioned critique by a scientist and a leader in the campaign against GM in England and around the world.

• Cohen, Robert, *Milk: The Deadly Poison* (Argus, 1998). Inside story of BGH, including the cozy relationship between the FDA and biotechnology.

• Shiva, Vandana, *Biopiracy: The Plunder of Nature and Knowledge* (South End Press, 1997). A Third World perspective calling for "collective intellectual property rights."

• Ticciati, Laura and Robin Ticcati, Ph.D., *Genetically Engineered Foods: Are They Safe? You Decide* (Keats, 1998). Manual by the founders of Mothers for Natural Law.

Organizations

• **Alliance for Bio-Integrity** has filed a lawsuit against the FDA to obtain mandatory labeling of GM foods. Alliance for BioIntegrity, Box 110, Iowa City IA 52244, (800) 549-2131. www.bio-integrity.org

• **Campaign for Food Safety** is a grass roots public interest group that maintains a web site devoted to genetic alteration of foods, organic standards, irradiation, and other issues. 860 Highway 61, Little Marais MN 55614, (218) 226-4164. www.purefood.org

• **Campaign to Label Genetically Engineered Foods** is a public advocacy group focusing on legislative action and citizens' initiatives. Box 55699, Seattle WA 98155; (425) 771-4049. www.thecampaign.org

• **Center for Ethics and Toxics,** Box 673, Gualala CA 95445; (707) 884-1846. Organization led by Dr. Marc Lappé and Britt Bailey, authors of *Against the Grain.*

• **Chefs Collaborative 2000** is a group of cooks and restaurateurs who oppose genetically altered foods. 25 First St., Cambridge MA 02141; (617) 621-3000.

• **Consumer Right to Know Campaign for Mandatory Labeling and Long-Term Testing of All Genetically Engineered Foods,** 500 Wilbrod St., Ottawa, ON Canada K1N 6N2; (613) 565-8517. www.natural-law.ca/genetic Clearinghouse for information, including a web news service, managed by Richard Wolfson, Ph.D.

• **EarthSave** is a national organization founded by John Robbins to promote a plant-based diet and food safety. The Boston chapter has been especially active on GM issues. www.bigfood.com/~EarthsaveBoston/

• **Eden Foods,** thepioneer natural foods manufacturer and distributor, has taken the lead in the campaign against genetic engineering. 701 Tecumseh Rd., Clinton MI 49236, (517) 456-7424. www.edenfoods.com

• **Edmunds Institute** is a nonprofit focusing on sustainable agriculture, biosafety, and biodiversity. It distributes tapes and literature related to genetically altered crops and foods. 20319-92nd Ave. W., Edmonds WA 90030, (425) 775-5383.

• **Genetic ID,** founded by John Fagan, pioneer scientific critic of genetic engineering, provides genetic testing for the food industry. 1760 Observatory Dr., Fairfield IO 52556; (515) 472-9979. www.genetic-id.com

• **Greenpeace International** coordinates worldwide education and nonviolent direct action on issues related to the environmental crisis. The GM site on its web page is: www.greenpeace.org/~geneng/

• **Institute of Science in Society** (ISIS) is devoted to creating a sustainable world. Its web site features statements on GM by scientists, especially Mae-Wan Ho and Angela Ryan. Flat 3, 42 Manor Rd., High Barnet EN5 2JJ, U.K. www.i-sis.dircon.co.uk

• **Kushi Institute** offers programs and classes in macrobiotic cooking and health care, including genetic engineering and other social issues. Box 7, Becket, MA 01223, (413) 623-5741. www.macrobiotics.org

• **Mothers for Natural Law** organizes a petition drive and distributes information. Box 1177, Fairfield IA 52556; (877) REAL-FOOD. www.safe-food.org

• **Physicians and Scientists for Responsible Application of Science and Technology** (PSRAST) organizes declarations on the impact of GM foods and sponsors a web site with scientific documentation. Tel +46-322622966. www.psrast.org

• **Physicians Committee for Responsible Medicine** (PCRP), led by Neal Barnard, M.D., supports a plant-based diet, animal welfare, and an end to BGH. 5100 Wisconsin Ave., NW, Suite 404, Washington DC 20016; (202) 686-2210. www.pcrm.org

• **Rural Advancement Foundation International (RAFI)** is a nonprofit monitoring Terminator Technology, gene patenting, and sustainable agriculture. 110 Osborne St.. #202, Winnipeg, MB R3L 1Y5, Canada. www.rafi.ca

• **Sierra Club,** the nation's largest environmental organization, has come out strongly against GM. 85 2nd St., San Francisco CA 94105; (415) 977-5600. www.sierraclub.org

• **Third World Network** focuses on issues of biotechnology and biosafety and features materials by Dr. Vandana Shiva, leading critic of genetic engineering in the developing world. 228 Macalister Rd., 10400 Penang, Malaysia. www.twnside.org.sg

• **Union of Concerned Scientists** mobilizes scientists and issues public statements. 2 Brattle St., Cambridge MA 02238; (617) 547-5552. www.ucsusa.org

Brave New World

• **The Monsanto Corporation,** the world's leading agricultural biotech company. 800 N. Lindbergh Blvd., St. Louis MO 63167, (314) 737-7642. www.monsanto.com

• **Delta and Pine Land Company,** 1 Cotton Row, Scott MS 38772, (601) 742-3351. Developers of "Terminator" technology and seeds.

Index